W0037573

CRITICAL GRAPHICACY

Science & Technology Education Library

VOLUME 26

SERIES EDITOR

Wolff-Michael Roth, *University of Victoria, BC, Canada*

FOUNDING EDITOR

Ken Tobin, *City University of New York, N.Y., USA*

EDITORIAL BOARD

Henry Brown-Acquay, *University College of Education of Winneba, Ghana*
Mariona Espinet, *Universitat Autonoma de Barcelona, Spain*
Gurol Irzik, *Bogazici University, Istanbul, Turkey*
Olugbemiro Jegede, *The Open University, Hong Kong*
Lilia Reyes Herrera, *Universidad Autónoma de Colombia, Bogota, Colombia*
Marrisa Rollnick, *College of Science, Johannesburg, South Africa*
Svein Sjøberg, *University of Oslo, Norway*
Hsiao-lin Tuan, *National Changhua University of Education, Taiwan*

SCOPE

The book series *Science & Technology Education Library* provides a publication forum for scholarship in science and technology education. It aims to publish innovative books which are at the forefront of the field. Monographs as well as collections of papers will be published.

The titles published in this series are listed at the end of this volume.

Critical Graphicacy
Understanding Visual Representation Practices in School Science

by

WOLFF-MICHAEL ROTH
University of Victoria, BC, Canada

LILIAN POZZER-ARDENGHI
University of Victoria, BC, Canada

and

JAE YOUNG HAN
Seoul National University, Korea

 Springer

A C.I.P. Catalogue record for this book is available from the Library of Congress.

ISBN-10 1-4020-3375-3 (HB)
ISBN-13 978-1-4020-3375-9 (HB)
ISBN-10 1-4020-3376-1 (e-book)
ISBN-13 978-1-4020-3376-6 (e-book)

Published by Springer,
P.O. Box 17, 3300 AA Dordrecht, The Netherlands.

www.springeronline.com

Printed on acid-free paper

All Rights Reserved
© 2005 Springer
No part of this work may be reproduced, stored in a retrieval system, or transmitted
in any form or by any means, electronic, mechanical, photocopying, microfilming, recording
or otherwise, without written permission from the Publisher, with the exception
of any material supplied specifically for the purpose of being entered
and executed on a computer system, for exclusive use by the purchaser of the work.

Printed in the Netherlands.

Contents

Preface

High school is dominated by textbook-oriented approaches to teaching and learning. Thus, a survey of 149 teachers revealed, biology students have to read, depending on academic level, between ten and thirty-six pages per week from their textbook.[1] One therefore has to ask, To what degree do textbooks introduce students to the literary practices of their domain? However, little research has addressed the quality of science curriculum materials, particularly textbooks, from a critical perspective. In this light, we are concerned in this book with coming to a better understanding of the reading and interpretation practices related to visual materials—here referred to as *inscriptions*—that accompany texts. Our overarching questions included: 'What practices are required for reading inscriptions?' and 'Do textbooks allow students to develop levels of graphicacy required to *critically* read scientific texts?' Some of the more specific questions included: 'What are the practices of relating inscriptions, captions, and main text?,' and 'What practices are required to read inscriptions in school textbooks?' That is, we are interested not only in understanding what it takes to interpret, read, and understand visual materials (i.e., inscriptions), but also in understanding what it takes to engage inscriptions in a *critical* way. It is only when citizens can critically engage with language (texts, speech) and inscriptions that they become knowledgeable users of television, newspapers, and magazines, who can choose or leave aside particular expressions as part of the particular politics that they participate in.

This book has arisen from a concern about teaching 'authentic science' to students. It soon had turned out that inscriptions, various graphical representations other than text play an important role in all disciplines for articulating theoretical models. However, we also noted that students do not receive, in general, specific assistance in practicing critical literacy with respect to graphical representation, a literacy that we refer to here as *critical graphicacy*. Over the years, we engaged in numerous projects to articulate the work of reading required in learning from graphical representations in science. Those studies,

vii

which appeared in journals such as the *Journal for Research in Science Teaching, Science Education, British Journal of Educational Research, Journal of Curriculum Studies*, and *Cybernetics & Human Knowing* constituted the starting points for the present effort. We reused some of the images, but reworked our analyses so that they (a) reflected our present-day understanding and (b) constituted a cohesive whole with the remainder of this book.

We acknowledge the support we received from different grants. Wolff-Michael Roth has received different grants from the Social Sciences and Humanities Research Council of Canada, which supported graduate students such as Lilian Pozzer-Ardenghi and research assistants who contributed to establishing our database. JaeYoung Han received a postdoctoral fellowship from the Korean Science and Engineering Foundation, which allowed him to spend one year in Victoria to conduct research under the supervision of Wolff-Michael Roth. We are grateful to both granting agencies for their support, but take full responsibility for the contents of this work.

We thank G. Michael Bowen and Michelle K. McGinn, who had worked as graduate student research assistants in the early phases of the research program on scientific inscriptions. All individuals in the CHAT@UVic research group during the 2003–04 academic year (Diego Ardenghi, Leanna Boyer, Damien Givry, Maria Inês Mafra Goulart, Michael Hoffmann, SungWon Hwang, Yew Jin Lee, and Giuliano dos Reis) allowed us to vet some of our ideas about representation practices, particularly with respect to chemistry. Robert Anthony has been a critical respondent to our work on photographs; our thanks are extended to him, too.

Wolff-Michael Roth
Lilian Pozzer-Ardenghi
JaeYoung Han
September 2004

Introduction

Every image embodies a way of seeing. Even a photograph. For photographs are not, as is often assumed, a mechanical record. Every time we look at a photograph, we are aware, however slightly, of the photographer selecting that sight from an infinity of other possible sights. This is true even in the most casual family snapshot. The photographer's way of seeing is reflected in his choice of subject. . . . Yet, although every image embodies a way of seeing, our perception or appreciation of an image depends also upon our own way of seeing.[1]

We live in visual cultures. Everyday print, computer, and television media inundate us with graphics other than text, including maps, charts, diagrams, tables, and graphs. All of these graphical forms other than text, frequently referred to as *inscriptions*, are used to integrate complex sets of information, to illustrate phenomena too difficult or cumbersome to describe in words, and to present data in succinct ways. In some situations, one figure, as the popular adage goes, can be 'worth a thousand words'. Yet, invisible and infrequently questioned, inscriptions embody selections and non-selections, ways of seeing and ways of non-seeing, inclusion and exclusion. Take the photograph of the San Juan Islands in Georgia Strait off Victoria, Canada (Figure 1). It shows a number of islands, clouds dramatically lit up by the setting sun. Photographs like this typically accompany the advertisements designed to attract tourists into this area of British Columbia. We use such photographs to show others—for example, our relatives living in Brazil, Germany, or Korea—how beautiful it is where we live. We also use such photographs, for example, on web sites or as email attachments to show people elsewhere in Canada how beautiful it is here during the months of March and early April, when the remainder of our country is still covered in snow.

But the photograph deceives without telling the observer so. There are, in John Berger's words that we chose for opening this book, many other possible sights at the same moment that this photograph was taken, much less advantageous, much less suited for bragging about where we live, much less suited to

Figure 1. Gorgeous evening in the Georgia Straits near Victoria, Canada, inscribed in a photograph taken from the plane by one of the authors during the rainy season.

advertise the area that we have made our (temporary) home. More so, a little later on the same day, fog and clouds pervaded the entire Georgia Straits, a weather pattern typical for the time of the year. That is, the photograph neither shows nor articulates what we may normally see and experience at this time of the year, clouds and fog associated with cool and damp weather—in fact, many Canadian friends told us they do not desire living on the West Coast. Knowing that 'rain forests' surround us confirms their suspicion that rain, clouds, and fog predominate our weather patterns.

We note that images deceive, because on their own, they do not tell us what they omit or on what basis the photographer chose to select them. Thus, as we are writing these pages, images of the 'war against terrorism' fill newspapers and television screens. Viewers who do not access the media of different countries will hardly notice what their own media present or do not present. Therefore, most Americans will not notice that their media present a rather one-sided view of what the war on terror is about, what and how the people of Afghanistan or Iraq think, or the suffering of innocent bystanders, to which the American army officers euphemistically refer to as 'collateral damage'.

There is more to it. Photographs such as figure 1 are used to say not only something about our area but also something about ourselves. We are what we say, gesture, write, represent in photographs, and do. The way we speak, gesture, write, and do are spoken, gestured, written, and done to mediate our development into the Selves we become. Through words, gestures, writing, and other actions, we build ourselves in a world that is building us. That world addresses us to produce the different identities we carry forward in life: men are addressed differently from women, aboriginal people differently from Europeans, 'A' students differently from 'D' students, or those coming from working-class families differently from those descending from nobilities. Yet, though means of communication are fateful in teaching us what kind of people to become and what kind of society to make, the way we represent and are represented in words, gestures, images, and actions is not destiny. We can redefine ourselves and remake society, if we choose, through alternative rhetoric and dissident projects.

Over the past decade, there have been an increasing number of studies documenting the pivotal roles and functions of representation practices in science. As a result of this work, representations other than text are called *inscriptions*.[2] They first appear in scientific laboratories and field sites, and—after having been cleaned, superposed, transformed—are later used in scientific publications, from where they are sometimes adopted into the popular media including Internet, magazines, and newspapers. Inscriptions include readings from simple devices, recordings from automated devices, computer screen output, photographs, micrographs, data tables, graphs, and equations. The more information an inscription summarizes, the more it becomes complex, resistant to deconstruction, and powerful. Despite the centrality of representation practices in science, relatively little work has been done in education from either sociological or psychological perspectives focusing on how people use, and above all, *critically* use inscriptions. Consequently, a deep understanding and sound theory of critical inscription practices does typically not inform the way inscriptions are used in education.

To get a handle on critical use of inscription, we use the notion of *graphicacy*, with which we refer to the knowledgeability relative to sketches, photographs, diagrams, maps, plans, charts, graphs and other non-textual, two-dimensional formats.[3] What is represented can be in an iconic relation with what we can see in the world, such as in photographs or naturalistic drawings, or have a more distant relation to the world, as for example, spatial maps, plans and diagrams, or can exhibit numerical relations (as in tables and graphs). Given the widespread use of graphics in newspapers, television, and instructional materials at the workplace, certain levels of graphicacy are presupposed in cultural life.

Textbook authors, too, presuppose certain levels of graphicacy among students and across subject matter domains, particularly in the sciences. Our research shows that there are between fifteen and twenty-four visuals used on every ten pages in science textbooks depending on country (Brazil, Canada, Korea), grade level, and subject area (biology, chemistry). It therefore comes as no surprise that educators have been calling for increased attention to students' competencies in reading various forms of representations other than text. That is, given the pervasive nature of images and inscriptions, verbal literacy and graphicacy are desirable competencies that students ought to develop to enable them to cope with these forms of communication as part of their everyday world.

However, literacy and graphicacy are not sufficient. In the research literature on literacy, the notion of *critical literacy* has been introduced to articulate particular forms of knowledgeability to be developed by students. Critical literacy allows for questioning power relations, discourses, and identities in a world not yet finished, just, or humane. From this perspective, literacy is understood as social action through language use that develops us as agents of change inside a larger culture, while critical literacy is understood as learning to read and write as part of the process of becoming conscious of one's experience as historically constructed within specific power relations. Here we propose the notion of *critical graphicacy* as a corresponding concept. Critical graphicacy allows questioning the power relations, discourses, and identities that human agents produce and reproduce using various forms of graphical representation. But why do we need *critical* graphicacy rather than simply graphicacy? In the following, we articulate some of the ideas and methods that led us to write this book.

Deconstructing indoctrination

Because students, as any human or organism more generally, adapt to the conditions in which they live, they adapt to the particular conditions of doing science, mathematics, or history without having to make thematic much of the context and activities in which they participate. That is, learning authentic science and other subjects would have some likeness with adaptation to particular settings and the structures that characterize its social and material characteristics. The way students thereby adapt is by means of developing structured dispositions. These dispositions generate patterned (i.e., structured) perceptions and with it the field of possible (material, discursive, etc.) patterned actions, that is, the practices characteristic of a field. However, it is the field that structures the dispositions as children learn to see the natural and social world in culturally characteristic ways. Because the structuring during enculturation goes unnoticed, acquiring dispositions is associated with acquiring the blind spots, ideologies, and prejudices of the field. Developing dispositions therefore means to acquire structured dispositions without noticing it. This is no different whether students acquire dispositions being taught in a traditional way or are taught by enculturation, for example, into the 'authentic' ways of science. In recent years, science has come under increasing scrutiny because, in many ways, it is not unlike traditional religions that requires their followers to believe in a basic, unquestioned and normally unquestionable set of presuppositions, the reigning ideology of the group. However, even within the different sub-fields of science, practitioners do not agree with all assumptions and question the very pedagogy used to bring new members into the culture. In a workshop for young practitioners of sociology, for example, Pierre Bourdieu advises the practice of 'radical doubt' with which his students are to get hold of and critique their own prejudices and commonsense notions that they had previously acquired without being aware of them.[4] One can therefore conclude that, unless science (mathematics, history, etc.) education includes a reflexive component that allows students to critically evaluate the knowledge claims of a particular field or media, it will always be subject to some form of indoctrination. Education for a critical graphicacy, paralleling education for critical literacy, means providing students with the opportunities to interrogate the different means of representing the world and to question the different power relations that are thereby constructed.

Indoctrination: an example

As an example of the unquestioned manner in which variables in graphs are presented as decontextualized facts—rather than as the result of protracted experiences and modification of practices until observation, discourse, and manipulations were consistent with each other—take the following case. In the recording of an introductory ecology lecture at a local university, there is little discussion about the nature of any variable and the purpose of its selection. Graphs are presented together with their axis labels: this implies that natural phenomena can be understood in terms of the cause-and-effect relationship or correlation displayed. To the students in the ecology course the reductionism embodied by such a treatment is invisible because the variables are accepted as real. One part of the resource lecture focused on substitutable resources and was accompanied by a graph said to display this kind of resource (Figure 2), the array of examples might well lead students to understand that there was a considerable number of possible resources to look at, but no opportunity to try to understand why those specific variables were chosen for discussion over any other possible variables.

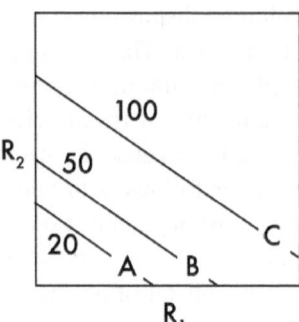

Figure 2. A graph used in an introductory ecology course to teach the relationship be-tween two independent variables (R_1, R_2) and an independent variable, here presented in terms of isoclines (lines of equal effect). Learning to understand graphs that represent the relationship between two independent and one dependent variable may lead to a par-ticular worldview, which holds that the world can be decomposed into small numbers of variables without loss of understanding a system as a whole.

The following excerpt from the lecture transcripts illustrates this argument as the professor provides a literal reading of the graph in Figure 2 depicting re-source interaction.

Substitutable resources, they are substitutable, either can replace others, they may not be equally good. So if you're a lion or cheetah, or on a savannah– Well if you need zebras or if you need gazelles, they are both pretty much as good. Similar to each other as re-sources [you can] either live on, over a year, thirty gazelles or maybe twenty zebras. And another example of this is when you're on an airplane and flight attendants come down the aisles and say, 'Chicken or fish, chicken or fish, chicken or fish'. Well they're substi-tutable in here, it may be better to say, 'They may not be equally bad'. This isocline shows substitutable resources. If you have a little bit of R-two and quite a bit of R-one the population will do about as well as if it has a little bit of R-one and quite a lot of R-two. So you can substitute one for the other. They may not be good but we can substitute.

This episode is rather typical for the way in which students are presented with variables (i.e., types of resources) as *the* joints along which the world *has to be* cut without any context of why these particular cuts were made. The re-source lecture presented students with a considerable number of 'examples' of the different outcomes from the interaction between two nutritional resources. Yet, even in the talk of resources that preceded it there was little mention of the variety of resources necessary for organisms to survive. The various examples

offered represent a bewildering range of categories, units of measure, and outcomes of effect (e.g., potassium, calcium, nitrogen, magnesium, pounds, kilograms, growth, growth rate, maintenance of health, etc.). There existed few opportunities for students to learn about the range of possible resources specific to
an organism and therefore they had little context within which to disclose and
develop their understanding of the relationship between two resources when reconstructing the larger picture that constitutes understanding a 'realistic' situation.

In this as in other lectures, variables are presented as a matter of fact; there
is little discussion about how and why any specific pair of resources matter in
the study. Our interviews with scientists showed that they usually need such
context to resituate the results of any one comparison within a broader understanding of ecology and resource use. Choice of which resources to be used for
study would not likely occur for arbitrary reasons and yet the choice of specific
variables for interaction studies is not a transparent process for newcomers to the
discipline such as the students in the course. The lack of descriptive context is
therefore particularly problematic for students in a class where they, for the first
time, learned about resources and their interactions and, also for the first time,
faced a two dimensional depiction of those interactions. Students were listening
to the talk about ecology as newcomers, without the insight that field experience
lends to the comments being made, listening to lecture talk that is structured as
if old-timers were telling their autobiography.

Working on the frontier of the objectified world known to the students in
the lectures, they search in the dark. Students work 'in the dark', as in traditional
laboratory exercises, apparently caught in a double bind. To know whether what
they observed is what they should have observed in the graph or picture, they
need to know that what they have done was what they should have done. It may
help readers to think about the following vernacular example. Novice cooks find
themselves in such double binds. If they are fortunate, their cookbook provides a
picture of the final result. Although they follow the steps of the recipe, they cannot know whether what they have done is what they should have done. However, many cookbooks have pictures that serve as check points to exit the double
bind. For, if their own final result is dissimilar to that displayed, they know that
what they have done was not what they should have done though they often do
not know exactly what they should have done differently.

At the same time, to know whether what they have done is what they should
have done they need to know that what they had observed is what they should
have observed. Our research in 'open inquiry' science classrooms and among
field ecologists shows that what are possible and viable ways of objectifying the

world only arises with experience in the particular domain or part of the natural world. Thus, variables are not just identified and perceived as stable structures in the world, imposing themselves on the observant scientist or science students. Rather, variables emerge from the coincident fit of natural constraints, manipulations of equipment and materials, and the (descriptive and theoretical) discourses humans use to objectify their experiences.

Our analyses of students' discourse surrounding graphs showed that they bought into the reductionist approach of much of Western science. For example, they spent considerable time and effort trying to place scientific inscriptions in an ecological context that would reflexively elaborate inscription and context. However, their discussions also suggested that they had insufficient experiential resources relative to ecology to aid their interpretations. Thus, engaging students in reflections on why *these* rather than other variables might be an important aspect of lectures designed to help them understand the 'reality' of some specified situation. Within ecological reasoning, the ability to move and re-cluster is an important aspect of whether the situation *is* biologically realistic or not. Furthermore, alternative ways of considering ecological systems such as those drawing on traditional ecological knowledge could contribute to enhancing students' critical stance towards all forms of ecological knowledge. At this point in time, school science does not provide students with the opportunity to be more reflective about graphs and other inscriptions. The students whom we recorded while attempting to interpret the graph did not have the interpretative resources to address any of these other variables. Teachers normally stop students' attempts to bring in other aspects of the world; these moves then serve to induct students to look at the world in reductionist ways.

Scientific representations (inscriptions) embody epistemologies

High school textbooks and scientific journal articles differ in the way they treat Cartesian graphs with data and those that feature models (see chapter 2); that is, there is a difference between (a) data which are used to ascertain or to induce some relationship and (b) graphs that express a relationship that was deduced from a theoretical model. In scientific publications, there is a recognizable relationship between 'real' data and the theory that is expressed in continuous line graphs; behavior is reduced to a single law or set of laws. The clean line of a smooth curve is given authority by the fact that the individual measurements do not fall on the line, and frequently are associated with bars that indicate errors of measurement. Points suggest accuracy and statistical variation, whereas the smooth curve approximating them holds out the hope for a simple relationship that can be expressed in a mathematical equation. Smooth curves displayed alone are looking for data points. Reviewers in the scientific community often take graphs without actual data as 'lazy attempts at demonstration'.[5] But such graphs are exactly what readers of high school textbooks continuously face. This interaction between empirical data and relationships of theoretical nature is no longer available to the readers of science textbook. Here, graphs are detached from empirical situations to which they might relate. But textbook authors never make it clear that the featured line graphs are used to express currently accepted models; this contributes to and constitutes the reigning ideology.

Scientific texts are concerned with nature; they provide us with ways of looking at nature. Scientific texts and the inscriptions they use serve as demonstrations in which readers are shown, and learn to see, the order reflexively specified by the textual arrangement of words and graphics. The text, then, is a work site where knowledge is constructed not through the passage of information from the book or article to the reader, but through the recovery of the order encoded by the author in text and accompanying graphs. It is through the noticeable work of reading that the knowledge of biological (natural) objects is constructed. The convincing text engages readers in a process of authentication so that readers recover what the biologist previously saw. Texts that achieve this authentication, that is, texts that achieve the demonstrable fit between graphs and text, display a compelling pedagogy in teaching readers something about the world they inhabit.

Recent curriculum reform efforts that focus on 'authentic' science, math, or history, have not sufficiently dealt with the problematic aspects of enculturation metaphor that comes with 'authenticity'. Because inscriptions are inherently partial and ideological, and therefore political, subject matter teaching should also include opportunities for students to critically interrogate the forms of inscriptions chosen in the domain. This would lead us to education concerned with 'epistemology as practice'. In such an education, students would not merely learn science, mathematics, or history, but as part of their activities of studying a domain also engage in an epistemological critique which would assist them in re-contextualizing and thereby relativizing all forms of knowing—both in terms of the opportunities they present and their shortcomings. Other work also shows that students' point of view relative to the ontology and epistemology of science changes when they concurrently read and discuss philosophical issues related to knowledge of a field and the way in which it is represented. We provide a brief glimpse into a physics classroom (taught by Wolff-Michael Roth), where students—in addition to conducting experiments and working with the textbook—read, wrote essays, and discussed readings on epistemology.

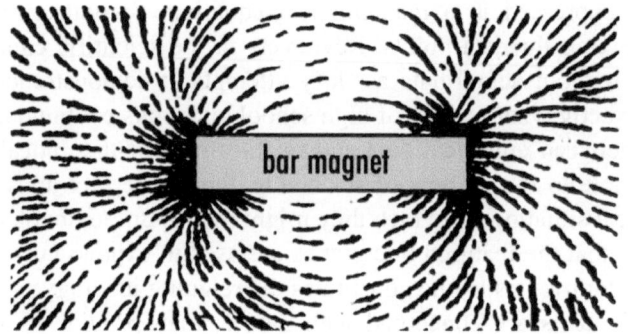

Figure 3. Learning to interrogate inscriptions, that is, the development of critical graphicacy *should be a core aspect of school curricula. The twelfth-grade physics students had
created this representation and now discussed what it was evidence of.*

Critical graphicacy in epistemological praxis

Many students never engage in epistemological discourse at any time during
their schooling. It is therefore not surprising that, during interviews, they appear
to have limited discursive repertoires for engaging in the topics broached by
educational researchers. At the same time, when students are provided with opportunities to talk about epistemological issues, they can develop quite sophisticated ways of getting at core issues in the domain. That is, they develop an epistemological practice rather than confronting the domain as a monumental body
of knowledge. Epistemological practice here means that students engage in
questioning visual re-presentation rather than trying to regurgitate what is presented to them as unquestionable facts. If such conversations occur as part of
science education (here physics), students are in fact provided with opportunities
to choose as a conversational topic the, until now unquestioned nature of scientific knowledge and the relation of this knowledge to its objects. Such questioning is exemplified in the following discussion that critically examines the nature
of magnetic fields.

01	Mark:	Magnetic fields are real. You can take a magnet put a sheet over it and pour iron filings on it, and you can actually see magnetic fields. ((Students use a magnet and iron filings from cupboard and produce a representation similar to Figure 3.))
02	James:	Not necessarily.
03	Teacher:	What *do* we see?

04	James:	We see the iron filings. We don't necessarily see the fields.
05	Todd:	We see a pattern. It is obvious–
06	James:	You don't– It could be anything, it doesn't have to be necessarily what we think.
07	Todd:	What happened when we came? We can actually see it– we can't actually see it.
08	Craig:	Who knows?
09	Peter:	I am just making it up. You guys don't exist, because I just made a major statement. Maybe I am all alone, you see? And to stop my loneliness, to stop me from going crazy I invented all of you guys, including physics class.

Here, started off by the teacher's question as to the ontology of notions such as 'energy' and 'field', students use as their point of departure the iron filings that had arranged themselves in a semi-circular pattern around the ends of a bar magnet (Figure 3). Students thereby take a specific case of an inscription about which they agree (patterns inscribed in the ordered iron filings) and use this as an occasion to argue whether this perception constitutes a proof of the existence of 'fields'. It appears as if the conversation generates the possibility of a discussion about realities. However, the conversation not only raises the possibility, but it also suggests relationships between these realities and language. In the process, brief questions by the teacher can pick out a particular aspect of the discourse and put it in relief. In the present case, 'What *do* we see?' encourages the conversation to focus on the perceptual act. As the conversation unfolds, different positions are discursively constituted. Whereas Mark's description states that fields are perceived (speaking turn 01), James suggests that it is not clear what they had seen beyond iron filings arranged in some noticeable pattern (turns 02, 04, 06).

The conversation allows students to work out different positions, and thereby, ultimately, identify themselves with one of them. The unfolding discussion feeds itself, lays down a garden path in walking, but also provides further grounds to be taken up later in the conversation. But almost out of the blue, Peter puts forward a radical proposition that raises the stakes: Reality is the result of making statements, is constructed by means of language. This changes the direction of the conversation's content to become a conversation about language. Peter thereby formulates the question as to whether discourse statements alone suffice to bring something into being. And he does so with a lot of humor by bringing forth the possibility that not only the conversation, or the conversation about the conversation, but also that the very setting in which he and his peers find themselves is only a figment of his imagination.

Thus, critical reflection on knowledge and its representation while studying a domain such as science (mathematics, history, etc.) provides students not only with the opportunity to participate, to some limited but hopefully legitimate peripheral degree, in the discourse of some field, but also in reflecting critically on this knowledge and its visual representation rather than uncritically accepting it. Thus, we would like to see a more reflective stance on the part of educators when it comes to constructing learning environments designed to introduce students to the knowledge of some domain.

Toward a critical graphicacy

We see an emancipatory power that epistemological practice confers to students, which influences our decision to take a stance against indoctrination while doing critical graphicacy as part of an epistemological praxis in school classrooms. The contextualization of inscriptions in a broader context of epistemology allows students to develop a relation to this knowledge that will free them from the authority of the expert. They will be able to take into account (a) the limited and contextual validity of expert advice or journalists' representations in the print and television medium, for instance, in the sociotechnical and environmental controversies that have become part and parcel of life in our societies and (b) the ways in which conflicts are represented. As citizens, they would then participate in referenda on science policy in ways similar to the Swiss citizens' participation in a referendum on aspects of the research in biotechnology. The Swiss people consider such participatory practice a normal and natural feature of their everyday lives. There is no reason why this would be any different in other countries, places, and communities. More so, in reconsidering the value of the different forms of knowledge and thereby giving a new status to common visually represented knowledge, students are also prepared to be open to considering different forms of knowledge coming from other cultural contexts. The ecological knowledge of the First Nations People, for example, would cease to be looked upon as some kind of folklore and given the status of authentic knowledge. At this point, epistemological practice is demystified by becoming part of everyday citizenry leading to tolerance in respect to other cultures and the forms of graphical knowledge they have developed. This tolerance becomes a source of enrichment of diversity and variety for individuals and collectivities. We can be hopeful, for as our excerpt illustrates, high school students can engage in epistemological practice, which includes critical graphicacy and literacy as its central components.

In the chapters that follow, we deconstruct different forms of inscriptions that are used in textbooks, and analyze the practices middle and high school students enact when asked to interpret inscriptions. To do the deconstructive work, we followed an approach whereby we investigated what is required for reading and understanding inscriptions through our own reading practices. Such an approach to reading, therefore, is aptly called phenomenological anthropology of inscriptions—anthropology, because we are concerned with practices, phenomenology, because the practices we study are those that we exhibit, being competent members in the field of science education. We are, in a sense, individuals that display considerable levels of graphicacy—in part because of our extensive research over the past decade. We therefore analyze graphicacy by means of *our* reading of inscriptions.

Toward a phenomenological anthropology of inscriptions

What counts in particular disciplinary contexts as convincing arguments or persuasive results is inexorably tied to the practical devices through which phenomena become *accountable*.[6]

Readers might ask, why investigate graphicacy in terms of practices rather than cognitive ability. Here, we briefly articulate our motivation to pursue a research agenda concerned with representation as practice rather than as something mental. We do not dispute that there are things going on in the minds of people who use and produce inscriptions, but we find it more interesting to investigate what users and producers actually make available to those that they interact with.

Graphs as mental representations

In the educational and psychological literature, there exist a number of reports that focus on students' use of inscriptions in conjunction with textual information.[7] Such research indicates that visual displays can be used to present data, illustrate abstract concepts, organize complex sets of information, facilitate the integration of new knowledge with existing knowledge, enhance information retention, mediate thinking processes, and improve problem solving. Despite the large range of possible instructional functions of inscriptions, educational practice often displays an ambivalent attitude; research provides little evidence that inscriptions are living up to their potential. Researchers generally agree that the field of cognition related to inscriptions is still in its infancy and many questions remain unanswered.[8]

Cognitively oriented studies theorize inscriptions as a composite of individual cognitive abilities and skills, and view them as (mental) representations. In this tradition, researchers focus on the objectifiable relationships among inscriptions, and between inscriptions and corresponding algebraic rules and situations, as well as the function of inscriptions in students' cognitive operations. Science education research that studies graphing from cognitive perspectives comes to conclusions similar to those found in educational and psychological research that studies inscriptions more generally: students fare poorly in their attempts to interpret graphs; and students have misconceptions, are confused, and take, for example, graphs as pictures. The reasons for these performances are often sought in terms of 'deficits' in cognitive abilities, cognitive development, reading ability, reading comprehension, cognitive processing, and so forth.

While research from cognitive perspectives provides some understandings about individuals' use of inscriptions, we argue that this research does not address the critical dimensions of teaching and learning related to the representation practices affiliated with graphs. We suggest that representation practices have first and foremost social functions; explanations of cognitive nature are sought only for those aspects not covered by social accounts. There are important implications for classroom practice when graphing is viewed as a social practice shared in communities of knowers and utilized in the context of making scientific arguments.

Using inscriptions as practice

Cognitive research on inscriptions has proven useful to some extent, but it does not take into account the relationship between people and their settings and practices; it also fails to consider inscriptions in relation to individuals' motives, goals of the inscription-related activity, and operations involved in inscriptions. Traditional and cognitively oriented studies have a disciplinary blind spot that makes them overlook the social, political, and rhetorical nature of graphs. Cognitively oriented studies are more concerned with the potential of inscriptions to retain more information rather than asking whether such information or the practices associated with inscriptions are worthwhile and related to (scientific) literacy in general. An anthropological perspective of inscription-related scientific practices is concerned with the public use, organization, and functions of these practices. Such a view regards learning not as a matter of having students acquire information (subject matter), metacognitive knowledge, and information processing skills. Rather, learning is indicated by the extent to which teaching allows students to participate in the practices of the domain they are to learn; and from our critical perspective, learning is also instantiated by the extent to which students reflect on these practices in a *critical* manner. Such research focuses on the public displays of practices—life forms—into which students are apprenticed and enculturated, and on the ways in which users can take a critical and reflexive stance with respect to the inscriptions they use. Such existing practices serve as models of what 'everyone else is doing', an important ingredient to the appropriation of new practices from the community in which students take part. We generally follow Bruno Latour's *Seventh Rule of Method* according to which researchers should seek recourse to cognitive factors only if there is something left unexplained after documenting the many ways through which inscriptions are gathered, combined, transformed, and tied together.[9] Thus, we do not exorcise cognitive approaches but seek to explore alternatives to traditional cognition for understanding critical graphicacy in the hope of finding explanations for the many questions that the cognitive approach has been unable to answer. That is, we focus on those aspects of graphing that constitute the blind spots of cognitive research on the topic.

Learning to use inscriptions involves acquiring practices. Traditionally, textbooks have been important resources in students' learning. In the historical separation of science and technology, and the corresponding separation of knowing from doing, textbooks have become the most important resource for science content and teaching. The way inscriptions are deployed in textbooks therefore plays an important role in the lived experience of students and in their associated appropriation of graphing practices in the course of schooling.

Anthropology of inscription use

Anthropology attempts to clarify what a community understands as its knowledge of the world and how that knowledge is embedded within the practices of that community. A culture (or subculture such as mathematics) is lived through, and known in and as, the minute concrete details of the practices that make it what it recognizably is. Within the culture of using inscriptions, the publicly accountable and accounted—for work constitutes the knowledge of its members. We choose an anthropological approach to unearth, in materially specific detail, the forms and functions of inscriptions in textual pedagogy.

As ethnographers working in a different culture, we expected as a result of our work to better understand our own practices. In this effort, we count on the natural analyzability of practical action. Keeping in line with our phenomenological approach (that also characterizes ethnomethodology), we attempt to clean away talk about reading inscriptions and to examine the practices of reading inscriptions themselves. We attempt to discover the practices through our own reading, but in a domain where we (at least some of us) do not know the subject

matter. The following analogy may clarify our point. Most readers are so familiar with driving a car that they no longer have to attend to driving and can interact with others, use the car phone, or eat and drink. However, they will experience considerable difficulties when they first drive in a country where the left-
right sense of road and vehicle are reversed. At this time, drivers begin to con-
sciously attend to changing the direction to look for other traffic, shift gears,
stay on the road, or pass other vehicles. That is, the nature of normal everyday
practices are revealed as drivers begin to attend to those actions that usually are
tacit and unavailable to reflection because of their ordinariness. Reading and interpreting inscriptions in an unfamiliar domain operates similarly by revealing
practices that we have come to take for granted.

Inscriptions can be viewed as tools used by practitioners to make phenomena visible. The embeddedness of practitioners' work within the practices of
their discipline allow the practitioner to see beyond those practices to the real-
worldly objects that animate the profession; that is, for experienced practitioners
graphs are transparent and unnoticed tools. For competently reading ecologists,
for example, inscriptions are used as a direct way to see what they are intended
to depict such as 'competitive interaction in blowflies', 'livestock-mediated tree
recruitment', or 'reproductive seasonality'. Because any graph belongs to a
complex of inscriptions and texts that make each of these concepts, newcomers
to ecology have a much more difficult task because of the unfamiliar aspects associated with the concepts. Inscriptions, then, are no longer transparent, but un-
ready-to-hand and obtrusive tools that draw all attention to themselves rather
than to the phenomena they are said to be about.

To reveal the practices associated with reading inscriptions, we need to
make the familiar practices unfamiliar; we accomplish this by reading in an area
that is not or only little familiar to us—we are physicist, biologist, and chemist,
respectively. In this book, we uncover competent inscription-related practices
through our own reading and interpreting of inscriptions. The feasibility of the
anthropological project lies in restoring to inscription use its concreteness as real
activity—the inspectable, cultural practices of a community. That is, we attempt
to recover a praxis-based validity of the natural scientific practices related to
reading inscriptions that are embedded in textual materials. Through our reading, we attempt to uncover the degree of fit between two contrasting classes of
organizational material: one class including the inscriptions of scientific demonstration, 'the other being the natural language-based explanation which operates
on the inscriptions by seeking to exhibit its relevancies, details, composition, or
conceptual consistencies'.[10] In our own effort, we find how 'reading' construes
the orderly sense of the inscription and surrounding text in order to disclose 'the

particular constraints which the text presents in terms of its ordering and pedagogic arrangements [that] will figure prominently as a specification of the order
of coherence which it provides to the "rational" visibility of the claimed object'.[11]

Overview

In chapter 1 we argue that in contrast to past educational research that views
graph-related activities in terms of mental ability, a conceptualization of graphing as a semiotic activity should be taken. This move provides a more viable account of individual experiences, familiarity, and socio-cultural factors during
graph reading and of errors committed by students and 'experts' alike. A model
of semiotic activity is presented that contains two elements: the process of parsing the perceptual field to construct relevant signs and the grounding of a sign
through the dialectic of sign-to-referent and referent-to-sign movements. The
model of semiotic activity also implies an integration of research on graphing
(and other sign related activities) into a more general concept of literacy.

In chapter 2, we provide answers to questions about (a) the practices required for reading graphs in high school textbooks and scientific journals and (b)
the role of high school textbooks in the appropriation of authentic scientific
graph-related practices. For our analyses, we selected five leading ecology-
related journals and six representative high school biology textbooks. To allow
more detailed analyses, an ontology of graphs was developed. Our fine-grained
analyses based on this ontology yielded qualitative differences between the uses
of graphs and associated captions and main text as they appeared in high school
textbooks and scientific journals. Scientific journals provided more resources to
facilitate graph reading and more elaborate descriptions and interpretations of
graphs than the high school textbooks.

In chapter 3, we investigate students' approaches to contextual word/story
problems. It shows that problems in school mathematics do not become contextual by embedding them in more descriptions of story situations. Rather, a problem is contextualized if the mathematical practices in which students engage are
integrated in a larger array of meaningful practices. During an ecology unit,
eighth-grade students are observed in two types of situations: (a) open-inquiry
field studies that included the production of convincing representations (inscriptions) to support findings and (b) word problems with stories and student-
produced data based on these field studies. Analyses of students' mathematical
practices show that word problems did not become more contextual, although
the story situations were very familiar to the students (describing an aspect of

their own activities) and although the inscription used to present information had been previously produced by one of the students. Students' inscription-oriented practices in word-problem situations contrasted with those during fieldwork.

Chapter 4 is the first of three chapters related to photographs. Photographs are a major aspect of high school science textbooks, which dominate classroom approaches to teaching and learning. It is surprising then that the function of photographs and their relation to captions and texts have not been the topic of analysis. We investigate the function and structure of photographs in high school science. Our motivating research question is, 'What can students learn from textbooks when they study photographs?' To answer this and several subordinate questions, we selected and analyzed four Brazilian biology textbooks. We focus on the use of photographs and the relation between them, various types of texts, and the subject matter presented. Our analysis reveals that the structural elements of text, caption, and photographs and the relations between them differ across the textbooks and at times even within the same book. This, of course, influences readers' interpretations of the photographs and changing their role in the text.

Little is known about how students make sense of and learn from photographs; even less is known about the different resources available for making sense of photographs when they appear in lectures. In chapter 5, the use of photographs during lectures and lecture-type situations is analyzed with respect to the meaning-making (semiotic) resources that speakers standing next to the projected photographs provided for understanding and learning from them. Our analysis identifies eight types of gesture/body orientations as semiotic resources that decrease the ambiguity inherent in photographs, and that have the potential to enhance the understanding of photographs and the scientific concepts embodied in them. We surmise that teachers can help their students learn to read and interpret photographs from lectures when they project them in such a way that it allows the use of gestures and body orientations as additional meaning-making resources.

Our research reveals the dialectical character of photographs: they simultaneously lack determinacy and exhibit an excess of meaning. In chapter 6, we attempt to understand how, under this condition, high school students interpret photographs that are accompanied by different amounts and types of co-text (caption, main text). The data for this study consists of video-recorded interviews with twelve Brazilian high school students. What students perceive is in part a function of the presence of caption and main text; these texts not only *describe* what can be seen but also *teach* students how to look at photographs.

High school students not only need to develop subject matter literacy but also a literacy concerning photographs to fully understand their textbooks.

Research suggests that students do not easily understand inscriptions. This may be due to the gap between inscriptions and the things in the world that they stand for, a gap that requires considerable interpretive work to be bridged by the reader. There have been suggestions that overlaying an experience-distant inscription with one that is closer to students' everyday experience will help students learn. In chapter 7, we investigate the function of layered inscriptions in Korean middle school science textbooks. We develop a semantic model that allows us to describe the work of reading and interpreting such layered inscriptions. This semantic model picks up where we leave off in chapter 1, further developing the work involved in structuring a textbook page such as to reveal the relevant sign to be interpreted. Our analyses of several layered inscriptions articulates the tremendous amount of work that needs to be done to establish the links between the different layers and between the inscriptions and the world familiar to the student. In addition, different functional relations in layered inscriptions require different kinds and amounts of linking work. We show that although layered inscriptions decrease the gaps between more experience-distant inscriptions and the world of experience, the total number of different types of work (structuring, transposing, and translating) to be done and aligned increases. This chapter therefore provides a framework for studying how students learn from using inscriptions in general and layered inscriptions in particular.

More than the textbooks of other countries, Korean chemistry texts include colorful inscriptions. How, we might ask, do such inscriptions help learners of chemistry? In chapter 8, we investigate the function and structure of chemical inscriptions in middle school science textbooks by drawing on a semiotic framework. We develop the concept of *chemisemiotics* to unveil the work of reading required to understand chemical inscriptions in the way their authors intended them to be understood. We assume that different kinds and functions (structure) of inscriptions constitute different signs in the learning process. We show that the difficulty in understanding the particulate nature of matter may result from the different processes of semiosis (interpretation and meaning-making) between inscriptions depicting macroscopic particles and models based on microscopic particles. Our investigation provides a guide for presenting chemical inscriptions in textbooks, and for understanding how students learn from chemical inscriptions.

Throughout this book, we analyze inscriptions that appear in textbooks in terms of the work required to read them according to the authors' intentions; we consider both simple and layered inscriptions. We also analyze the graphicacy that students of various ages (eighth through eleventh grade) exhibit when they read, interpret, and transform inscriptions. In chapter 9, we take a look at the opportunities and constraints computers provide users to exhibit graphicacy, that is, manipulate inscriptions, which are then animated by the medium for exploring the consequences of the manipulation. In chapter 9, we develop issues in the semiotic and anthropological analysis of inscriptions that are articulated in previous chapters: layered inscriptions (chapters 7, 8) and the salience of signs with movement and gesture during face-to-face interactions (chapter 5). That is, we provide analyses of (a) the inscriptions that appear as part of a software program used in physics courses as a means to model movement phenomena and (b) student-student and student-teacher interactions over and about the inscriptions and events on the computer screen.

Throughout this book, we present analyses (a) of the work of reading required for understanding scientific inscriptions in different domains and (b) of students actually engaged in reading such inscriptions. However, our aim as critical educators is not just the provision of opportunities for students to become graphically literate; rather, we want students to develop *critical* graphicacy, that is, literate in constructing and deconstructing inscriptions, the deployment of which is always inherently political. In the Epilogue, we elaborate some of the steps we might want to take in the project of assisting students to develop not just graphicacy but *critical* graphicacy. We begin by providing a brief case study from our research that exhibits what we have in mind when writing about *critical* graphicacy. We then argue that educators ought to take an anthropological perspective on learning, which allows them to conceive of *critical* graphicacy as practice, inherently public and shared, rather than as stuff in individual minds. We provide an example from a physics course taught by one of us, which shows what critical educators might want to do in their own classrooms to develop *critical* graphicacy. We close with recommendations for the redesign of inscriptions and for issues critical educators might want to consider in their own teaching.

1   Toward a critical graphicacy

The (post-) modern world is visual. Relations between visual expressions and thought are so profound that they are no longer salient in conscious thought. Historical studies revealed that the very rise of science as a dominant cultural paradigm coincides with the use of graphical inscriptions.[1] Over the past two decades, a flurry of sociological studies showed just how central graphical inscriptions are to scientific laboratory work and communication, and to the co-ordination of collaborative work in science and technology.[2] It is therefore not surprising that educational researchers and educational reform documents call for an increased attention to the use of graphical inscriptions in education. These calls are intended to bring about learning environments that help students become literate in practices related to the reading, production, and use of graphical inscription. Interestingly enough, although mathematics and science educators speak and write about literacy (ability to participate in the discourse of a domain), a corresponding move to theorize graph use in terms of literary theory has not been made. Yet, when viewed within a more general framework of signing practice, graphing becomes one aspect of a more general concept of literacy that renders problematic the structure and 'practice of representations as a means of organizing, inscribing, and containing meaning'.[3]

In this chapter, we contribute toward an understanding of *critical graphicacy* by showing how reading one type of inscription, Cartesian line graphs, can be conceptualized as semiotic activity, which subsequently allows us to think about critical graphicacy more broadly. We begin by providing examples of scientists and students who enact competent and problematic graph readings and then propose a framework that ties graph reading to a more general concept of literacy. This framework—grounded in semiotics, anthropology, and critical phenomenological hermeneutics—is used as a lens to re-examine existing research on reading graphs and as a platform to argue for a more integrated concept of literacy concerned with inscription practices across the curriculum.

Reading inscriptions and texts

Research on inscriptions has largely been conducted from two theoretical bases. In the first approach, the quality of graphs and individuals' abilities are the onto-logical starting points leading to particular attributions. In much of the tradi-tional research on graphical displays, the quality of inscriptions is considered as a property inherent in the display[4] and knowledge as a capacity situated in the mind.[5] Howard Wainer goes as far as suggesting 'that the ability to understand graphically presented material is hard-wired in'.[6] In the second approach, soci-ologists and anthropologists interested in scientists' uses of inscriptions note that graphing can be described as a social practice that gathers, combines, ties to-gether and sends back inscriptions. Bruno Latour therefore suggests as a meth-odological rule to focus on social and pragmatic issues of graph use; 'only if there is something unexplained [. . .] shall we speak of cognitive factors'.[7] This view is consistent with pragmatic and postmodern approaches where inscrip-tions and texts—i.e., signs and sign complexes—do not have meaning a priori; rather, meaning arises from the contexts of sign use and is always open to recon-figuration.[8] The first approach largely focuses on cognition located in individual minds while leaving out the mediational role of culture and individual experi-ence; the second approach focuses on the role of culture in inscription activity, and largely neglects the individual as experiencing being.

In this book, we take a strong social-semiotic position by invoking cogni-tive factors only if the social pragmatic approach leaves something unexplained. Before presenting a model that accounts for individual acts of reading graphs, personal experiences of the reader, and the reader's cultural context, we provide concrete examples from our extensive research on the use and interpretation of inscriptions. For heuristic purposes and to facilitate succinct presentation, we categorize inscription-reading performances as falling onto a continuum includ-ing 'transparent', 'competent', and 'problematic' readings of graphs; but we be-gin with an example of reading verbal text that serves as a contrast for the sub-sequent examples of reading graphs.

Reading text, transparently and otherwise

An individual reading a text attempts to find out what the graph (qua text) says about some aspect of the world. This activity occurs in the context of an indi-vidual's familiarity with the signs (words, concepts), the 'natural objects' that the signs stand for, and the social conventions that regulate and constrain the use of signs. Thus, when signs and the things they refer us to are very familiar, read-ing leaps beyond the material basis of the text (a matrix of signs) to the things the text is said to be about. For example, competent readers of English have no trouble knowing what the following paragraph from a community newspaper is about.

As portables are removed from the schoolyards throughout the province in favour of building new classrooms, some school trustees are questioning whether the money couldn't be better spent.[9]

As the paragraph is read, the eyes do not even stop to identify individual letters or words; the sentence is not structured, but reading leaps beyond the material basis of the text to a reality that seems to lie behind.[10] Habitual processes of reading allow the competent reader to get to the issues at hand: schools, school yards, portables, the continuing financial crises in education, questions about how to best spent the available money, and so on. One can even imagine participating in the public debate for the purchase of books and computers instead of the construction of classrooms. Our experience in the realm of education provides us with the background to know what the text is about. In other words, the text becomes transparent to the activity of reading, seemingly allowing direct access to the phenomena that the text is said to be about. Thus, understanding a text involves knowing signs and having experiences in the domain that these signs refer to. If readers do not have experiences in, and familiarity with the domain a text refers to, that is, with the domain of empirical content, they are likely to have difficulties with the text. We exemplify this with a quote from a book that makes an argument about the empirical content of sentences (qua text), that is, the very topic of the present paragraph.

The following text is more difficult to understand, despite the fact that it is in English and makes use of two sign elements ('{}', 'q $\supset$ c') with which one might expect a college student to be familiar with because it is the topic of introductory courses on philosophy, mathematics, or logic.

Any sentence, even Russell's 'Quadruplicity drinks procrastination', is a supporting member of a set that implies an observation categorical. Let us abbreviate Russell's sentences as 'q', and some observation categorical as 'c'. The two-member set {'q', 'q $\supset$ c'} implies 'c', but the one-member set {'q $\supset$ c'} does not. So Russell's sentence is a supporting member of {'q', 'q $\supset$ c'}.[11]

At issue in this statement is whether and how the empirical content of sentences—its semiotic referents—can be checked. Willard Van Orman Quine shows that, if a set *implies* a generalization 'c' ('observation categorical') then it is supported even by a sentence such as 'Quadruplicity drinks procrastination'.

Here, because our readers may be unfamiliar with the content of signs in this context, they may not understand what the sentence is *about*, despite the likely familiarity with the signs themselves. The problems of reading are compounded even further when readers are less familiar with the forms of inscription. For most people, mathematical formulas, musical scores, twentieth century art, and graphs are not in common everyday use in the same way that verbal texts are. Thus, we might expect students and adult readers to exhibit difficulties with reading inscriptions such as mathematical formulas, musical scores, twentieth century art, and graphs not because of 'misconceptions' or 'cognitive deficiencies' but because they are unfamiliar with the content domain and conventions regulating sign use in it. We therefore propose to conceptualize reading inscriptions as but one example of a more general activity of reading signs. Whereas other educators have also made use of semiotic frameworks, they generally have not attended to those cases where the nature of the sign itself is in question.[12] We therefore propose a semiotic framework that includes two processes heretofore not articulated by other researchers. First, there is a process in which the individual has to perceptually *structure* the visual field to construct the sign itself as well as the interpretation.[13] Second, the sign is part of a dialectic process of sign-to-referent and referent-to-sign movements that mutually stabilize each other and thereby establish a corresponding referential ground for the sign. (Semiotic research generally considers only sign-to-referent interpretative processes.) These processes are exemplified in the following examples featuring 'transparent', 'competent', and 'problematic' readings.

Transparent reading

When scientists and other people, like the water technician in the following example, are thoroughly familiar with some graph and the part of the world it refers to, their readings become transparent in the same way that the text about portables in schoolyards does. Individuals look at graphs and, without hesitation, see in each wiggle—or change in height, change in slope, width of a peak, and so on—a corresponding state in the world. In fact, it is overstated to say that these individuals 'see [something] *in*' a sign. Often, the scientists, engineers, and technicians in our studies do not even point to specific features of a graph; they simply describe a situation and expect listeners to see in the graph what they are talking about. Familiar graphs provide them with a transparent window onto a familiar world, in the same way that a paragraph about portables, schoolyards, trustees, and finances provided a look at a familiar school-related world. Thus, in one instance, a biologist briefly explains an experiment on the feeding behavior of three types of frog larvae in the presence and absence of a caged predator.

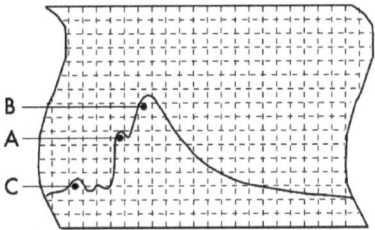

Figure 1.1. This graph displays the water levels in a creek. Just after rainfall events, the levels rise. As part of an environmental open-house event, Kelly points to different parts of the graph but talks about water coming from the north arm of Hagan Creek and off Mount Newton or from the south arm of the creek. She made no distinction between the graph and the natural phenomena she talked about.

Without transition, and without highlighting a particular feature of his graphs, he then reads the graphs as a text about the interaction of food availability and predation on the mortality of frogs. He did not have to structure the graph, and the graph was deeply grounded in the phenomenon. For this biologist, the graph and the phenomenon had fused.

   In a similar way, other scientists read from their graphs non-obvious phenomena. Our research shows that scientists are not the only ones who read graphs transparently. In the following excerpt, Kelly—a water technician whom we recorded on four different occasions (at work and during a public display)—transparently read a graph (figure 1.1) produced by a water flow monitoring station at Woodwyn Farms: ([A], [B], and [C] are locations on the graph to which Kelly pointed as part of her talk [figure 1.1].)

This peak has an earlier tiny peak ((points to [A])) and then the main one ((points to [B])). And from the data collection that I've done, we know that the north arm [of the creek] has approximately twenty percent of the volume, it comes off Mount Newton. And the south arm has about eighty percent of the total volume up to the point where they meet. . . . So, what the blip ((points to [A])) means, then, is that when we get a rainfall, we had quite a bit there, the north arm, the water from the north arm comes off faster. It's less volume but we see the peak ((points to [A])) first, down at the Ocean Farm station. Then it takes quite a bit longer, 8 hours perhaps to show that ((points to [B])) rainfall.

   In this excerpt, Kelly read the graph transparently, the source of the water passing her monitoring station at different moments and the time delay between water from different arms of the creek. Noticeably, Kelly read one wiggle [A] as a sign that water came from the north arm of the creek but did not consider as significant similar wiggles [C] visible to the left of the peak (figure 1.1). In scientific terms, she read some wiggles [A] as signals (i.e., as something *signi*-fying or *signi*-ficant) and others [C] as noise (i.e., as something in-*signi*-ficant).

In the same way, Kelly read from the graph at what point Woodwyn Farms dammed the creek to build up a summer reservoir, or at what point farmers in the valley turned on their irrigation systems. She pointed to a wiggle and said, 'This is a clogged pipe'[14], or identified diurnal changes in freezing (night) and thawing temperatures (day). She 'saw' that there was not enough water for the fish to survive, or that the farms had to stop irrigation because of lack of water. But Kelly also moved in the reverse direction, from the world she knew to the world of expressions (chart). She talked about local phenomena and translated them into the terms of the chart. She talked about the historical context of the community and then indicated how the chart would have looked like thirty or fifty years earlier. Thus, in Kelly's use, the chart was fully transparent. In her activities, the graphs were as transparent as the report on the portables is to educators and readers of this journal. Graphs and language, in common use, are entirely transparent to users' activities in the world.

Several other research studies have shown similar effects related to the use of graphs. Graphical displays serve physicists as a locus where they and their physical phenomena can come into symbolic contact; and intensive simultaneous interaction of individuals with motion graphs and the moving objects represented leads to a fusion of signs and their referents. Therefore, when readers are very familiar with a sign system and the things it refers to, signs themselves become transparent. Readers no longer think of words or parts of a line curve, but go directly to the things they know them to be about. This transparency is so pronounced that we forget the distinction between sign and referent; we confuse the map with the territory. A graph simply provides the material ground that organizes competent reading; but the graph also requires competent reading to be understood and a familiarity with the situations or type of situations to which the graph refers. It is in that very disappearance of the sign, the leap beyond the material basis of the text, that reading achieves its social character.[15]

Competent reading

When texts (including graph, caption, main text) and their topics are less familiar, more work is required on the part of the reader. This work may involve an active structuring of the text (we often do this through highlighting, underlining, summarizing); the results of this structuring subsequently may be grounded in the world. If the nature of sign and phenomenon are uncertain, one may expect some dialectic process that fixes one particular reading of sign and phenomenon. Imagine Sherlock Homes at work, reading a situation, hypothesizing a series of events, re-reading, re-hypothesizing, and so on until his reading is consistent with a reasonable series of events. When he begins, even the nature of the sign is in question; it is only when the position of the chair with respect to the table becomes significant to him that he has constructed the sign that will lead him to the murderer. Such a dialectic process—in which the sign and inferred events mutually presuppose and stabilize each other—is just what we observed in scientists' reading of graphs.[16] Scientists elaborated descriptions of possible situations to which the graph (or part of it) possibly referred or translated descriptions of known and familiar situations in order to test whether these situations or descriptions were consistent with their reading of the graph. In this dialectic movement, then, scientists reified both the signs that made the text (result of perceptual structuring) and a phenomenon (result of grounding). In the following example, an ecologist read the left intersection of the population graph (figure 1.2.a) as an indication of a population in equilibrium. ([P$_1$] and [P$_2$] are locations to which the scientist pointed.)

So, anything below this point ((points to [P$_1$])), death rate is greater than the birthrate so that means the population size has to go this way ((gestures to the left of [P$_1$])) which leads to extinction. Anything above this point ((points to [P$_1$])) means that you have birth greater than death, the population grows to the point (points to [P$_2$]) where birth and death equal each other again.

Referring to figure 1.2.a, the scientist first pointed out that for population sizes smaller than that at [P$_1$], the population would decrease even further because the death rate (line) exceeded the birth rate (curve). Then he suggested that for population sizes larger than that at [P$_1$], the population would increase because the death rate (line) was lower than the birthrate (curve). After doing the same with the other intersection, he proceeded to test whether the graph describes the interaction between wolves and moose with which he was thoroughly familiar. Sketching lines as he went along (figure 1.2.b), he stated:

The rate of mortality due to wolves goes up as the number of prey. So this is N ((writes 'N')), meaning prey, so this is the mortality rate due to wolves ((begins to draw curve on the upward slope)), it goes up and then goes down ((finishes drawing on downward slope)) because the wolves are territorial and there's a limit and there's a limit on how many they take. So, the rate of death declines with population size eventually.

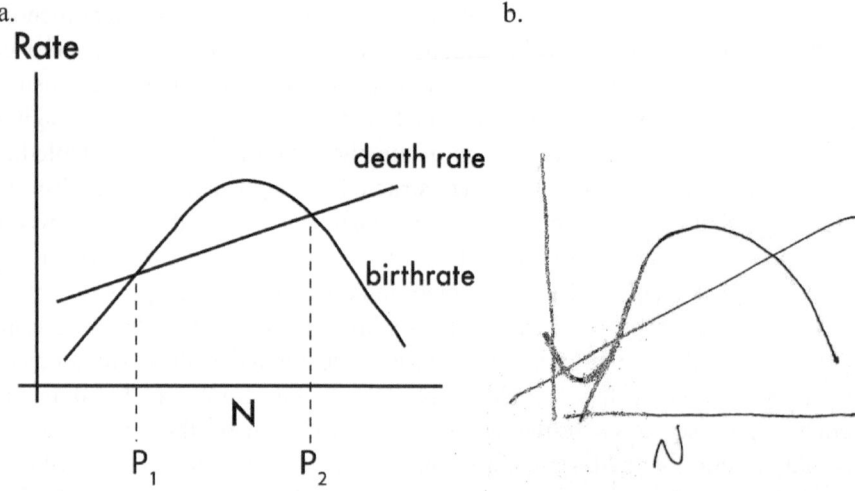

Figure 1.2. In the attempt to interpret the population graph (a), a research scientist also produces, as one interpretant, a second population graph (b).

In these excerpts, the scientist first provided a reading of the original graph (figure 1.2.a), then immediately moved to a situation he was very familiar with, the relationship between wolves and moose ('prey'). He then projected what a graph for this predator-prey pair would look like (figure 1.2.b) and compared it to the graph in the task. Readers can see from this and the previous example that reading a graph transparently not only requires readers' familiarity with reading graphs, the conventions regulating each of its component signs, and the correct identification of requisite component signs. It also requires a familiarity with the natural or hypothetical phenomena that such graphs may express.[17] Our research shows that in all those situations where students are familiar with situations described in the graph and with translating between phenomena and graphs, they experience little difficulty and do not get stuck. Their readings leap beyond the graph to a world that it might describe. On the other hand, even college science graduates experience difficulties reading graphs, consistently coincides with the observation that they are at once unfamiliar with translating between graphs and phenomena and with the natural phenomena themselves.

Competent reading of graphs is not age-dependent. Young students develop competent graph-related reading practices when provided with appropriate opportunities. In one of our research projects more closely described in chapter 3,

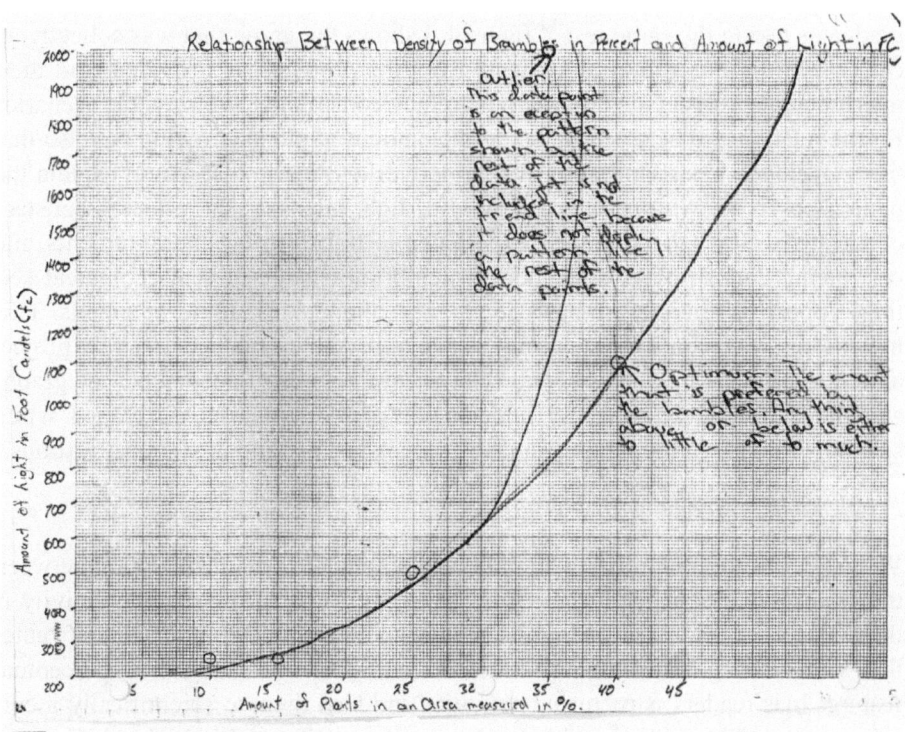

Figure 1.3. In their attempt to interpret a set of data presented on a map, grade eight students generated the graph above and alternative possibilities of interpreting it, based on different assumptions about particular data points.

eighth-grade students conducted their own research on the biological and physical factors affecting various ecozones. The main requirement imposed by the teacher was that students had to present their research projects and its findings in *convincing* ways and defend what they had done in regular mini-seminars with their peers. These students therefore had gained extensive experience in using mathematical inscriptions (including graphs) for their own intentions and in the service of convincing peers of the importance and quality of their research. The students also had extensive experience in translating phenomena into multiple mathematical inscriptions.

After ten weeks of working in this way, students were tested. These tests showed that the students, scaffolded by their scientist-teacher, had become more competent in data analysis, transformation, and interpretation than the B.Sc.

graduates taking the same test.[18] figure 1.3 shows the graph that was collectively constructed by one group of students from the problem set provided, and their interpretations. It is evident not only that they constructed two possible scenarios for the relationship between light and the amount of plants growing but also that they developed a description of a possible-world scenario that would explain the pattern observed. At a certain threshold of light intensity, the effect is constant or negative. Here, the students' experience in transforming measurements into convincing data displays, and their in-depth knowledge of real ecological settings appear to provide the basis for highly competent reading of graphs. This hypothesis was confirmed during a two-year ethnographic research project among ecologists that showed the relationship between transparency of graphs and the extensive background understanding the ecologists developed while spending months in the field capturing animals and making in situ observations.

Problematic readings of graphs

When people are unfamiliar with graphs, phenomena, or the translation between the two, problems in reading become apparent. Most of the reading activity is then concerned with structuring the graph (and accompanying text) itself rather than with relating it to some phenomenon. Iconic relations and salient perceptual features bias readers' structuring activities and lead them to scientifically incorrect readings. The following three examples exemplify the kind of problems that readers face.

Our first example concerns a standard problem used in the literature on graphing, which requires students to identify the faster of two vehicles given a distance-time graph; many students fail to provide the correct answer.[19] Thus, when asked to identify the faster vehicle at one second in figure 1.4, students typically respond with 'B'. A qualitative comparative reading of lines 'A' and 'B' shows that 'A' covers more distance in a specified time interval than 'B' and therefore is faster for the entire episode represented in the graph. The wrong answer 'B' is characteristically categorized as height-slope confusion because the answer to the question has to be inferred from the slopes rather than the height of the lines at the point of interest. Incorrect answers are typically based on height as the referent. The literature abounds with deficit characterization of students' performances on this and similar tasks. Thus, 'misconception', 'low-level thinking' and '[lack of] mental structures', 'deficient understanding', and 'deficient logical thinking abilities' have been used as explanatory resources. Researchers suggest that students fall prey to perceptually salient features rather than attending to the logical properties inherent in the inscription. In the semiotic perspective developed here, the relationship between signifier and signified

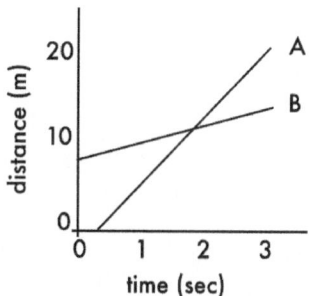

Figure 1.4. Type of graph frequently used to identify students' 'misconceptions' about graphing. A travels a larger distance for a given time interval than B and therefore is faster. Many students, when asked to identify the faster car at a time to the left of the intersection will incorrectly answer 'B'.

is always arbitrary, and the grammar regulating sign relations is based on conventions rather than being 'logical' a priori. Therefore, deficit characterizations may have to be revised given that, as the following two examples show, readings of the same type can also be observed among practicing and award-receiving scientists of international renown.

Our second example features a successful theoretical physicist and veteran professor who had volunteered to teach a course designed especially for elementary education majors. In this course, as in most physics courses that we have seen in our experience, the chalkboard was continuously 'littered' with inscriptions (figure 1.5), which requires tremendous levels of graphicacy on the students' part to come to an understanding of the lectures. Prior to this episode, he had released a ball on an incline, measured the distances covered after one, two, three, and four seconds, and recorded results in a data table. He first drew a distance-time graph (figure 1.6.a), and subsequently showed how the velocity-time (figure 1.6.b) and acceleration-time graphs (figure 1.6.c) could be developed from the first table and graph. To help students understand, the professor then introduced a money analogy from everyday life. However, as he used the vernacular example, his reading went awry (without him noticing it). Pointing to the acceleration graph, the professor repeatedly suggested during the lecture, 'Here I am not getting a raise– acceleration is staying constant. You get paid the same every day. That's what it looks like when one works for a university or for the state'. Here, his reading of 'staying constant' was translated into 'I am not getting a raise'. A careful analysis of the analogy shows that the three graphs

Figure 1.5. The chalkboard in the physics classroom was always 'littered' with inscriptions, requiring tremendous levels of graphicacy on the students' part to follow what the lectures were all about.

(figure 1.6) map onto total income (e.g. $) vs. time, wages (e.g. $/hr) vs. time, and change in wages ($/hr/yr) vs. time. Thus, a constant acceleration larger than zero indicates that the hourly wages increase constantly (e.g. $0.20/hr per year) rather than staying the same as the professor told his students. Despite his twenty-year experience, despite his success as an academic physicist and teacher, and despite being familiar with graphing and with income and salary, he incorrectly mapped the graphs onto the vernacular situation. Similar problems have been reported in different studies when professors attempted to link graphs to spontaneously developed vernacular situations. Incidentally, the students in this class never developed a sufficient level of graphicacy to read and critically examine the graphs on the chalkboard, homework, or exams, which let them frustrated and ultimately short of understanding the physics involved.

Our final example comes from a study of graph reading among scientists, who had been asked to respond to a variety of graphing tasks that undergraduate students in their own discipline would face in introductory courses (figure 1.7). An overall survey showed that between twenty-five and fifty percent of the scientists, who were mostly ecologists and biologists, incorrectly interpreted graphs (when compared to the accepted standard in the field) taken from a second-year university introductory course on ecology. An even higher portion of the scientists (ninety-two percent) responded inappropriately when asked to identify the

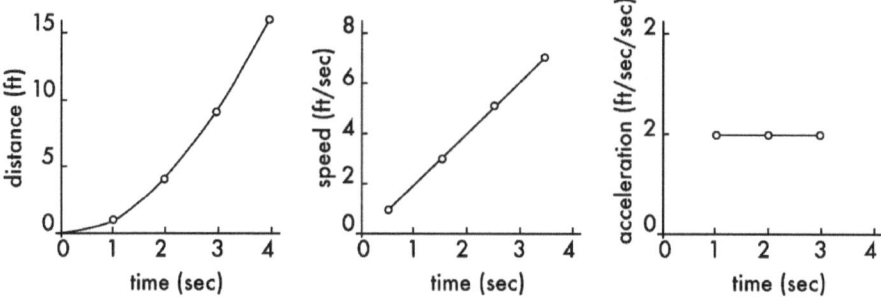

Figure 1.6. Series of graphs used in a physics course for preservice elementary teachers. When the professor provided a salary analogy, he inappropriately mapped (translated) the graphs into students' vernacular experience.

maximum increase of individuals in a population given the birthrate and death rate graphs (figure 1.2.a). About one-half of the scientists suggested that the population increased most where the birthrate was highest (see scientist in figure 1.7); the other half proposed that the largest increase was when the distance between birth rate and death rate was largest. However, because the graph represents the birthrates and death rates rather than the absolute numbers, the maximum change in individuals is where (birthrate - death rate)·N has its maximum. That is—similar to the middle and high school students in educational research who committed slope/height confusions—the predominant number of experienced research scientists in our studies used a perceptually salient feature as a resource in their (incorrect) answers. A parallel argument to the literature on students' graphing errors would suggest that the scientists fell prey to the perceptually salient height or to the height differences rather than inferring the conceptually appropriate answer.

   All three examples from our research show that scientists are prone to read graphs in ways structurally similar to, and equally incorrect from a mathematical perspective as those identified among students (first example). However, almost all of the scientists in this research were highly educated and well published and had received national and international awards for their research. In the light of these findings, it is difficult to uphold 'misconceptions' and 'cognitive deficits' as viable explanatory constructs. Rather than taking a deficit approach, we therefore propose a framework that moves educators away from the sole focus on mental activity.

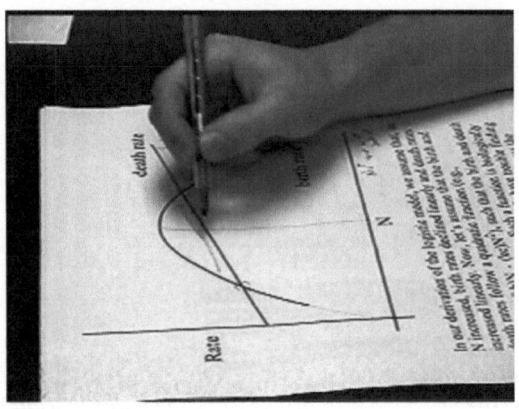

Figure 1.7. A scientist engages in the work of reading the population graph (Figure 1.2.a). He marks up the graph, thereby enhancing this graph and creating new inscription.

Reading inscriptions as semiotic activity

Educators can benefit a great deal and gain a lot of explanatory power in developmental theories by considering the semiotic aspects of the experimental tasks. Most educational research in semiotics is not concerned with the ontogenesis of sign practices. However, any theory of reading has to be able to explain the developmental trajectory of the reader. This trajectory begins with her initial perceptual parsing of the text to the transparent reading that comes with experience in and familiarity with the expressive domain (inscription, verbal text), experiential domain (world of phenomena), translation between the two, and conventions regulating the elements from the expressive domain. The conceptualization of inscription practices presented here is grounded in Peircean semiotics and hermeneutic phenomenology but extends this work in that it has a developmental component.[20] In this approach, inscriptions—including labels and numbers in the display, caption, and main text—are taken as texts, sign complexes, composed of signs. 'Interpreting inscriptions' is therefore subsumed into a more general activity of reading signs.

Signs and semiosis

The examples in the previous section show that even scientists may experience difficulties in interpreting unfamiliar inscriptions, and apparently make inferences based on perceptual rather than conceptual grounds. When scientists use vernacular examples for which they have not over-learned the translation to inscriptions, they also make errors. Furthermore, even highly trained individuals may find themselves in situations where they get stuck in perceptually dissecting a graph without ever connecting it to some external referent. On the other hand, these examples also show that people of all ages and training can read inscriptions transparently, that is, they treat these inscriptions as if they provided direct access to some phenomenon or situation. Such transparent readings always are the result of extensive experience with the sign system, phenomena, back and forth translations between sign and phenomenon, and conventions regulating sign use. A deficit model does not satisfactorily explain these research findings. Figure 1.8 displays the alternative model that we foreshadowed earlier: it includes perceptual parsing that constructs the sign and the dialectic grounding process in which sign and referent are stabilized.

The model consists of two major parts. The left-hand side represents the process by which readers structure the (graphical, verbal) text into signifying elements, signs. The results of this process become the input to the second process that grounds signs in a content domain (or referent). The general process of reading, and the two specific processes in our model are elaborated in the following paragraphs.

The key notions in this model are those of reading as semiosis laid out by Charles Sanders Peirce: sign (S), referent (R), and interpretant (I).[21] Signs are material traces (e.g. letters, words, texts, photos, or graphs) that refer the reader to something other than themselves, the referent R or content of the sign. For example, 'elk', 'birth rate', and 'population' are signs (here constituted by traces of ink on white paper) that signify, to an ecologist working on the elk population in Banff National Park, general natural objects and phenomena that she has experience with on a daily basis. Interpretants are commentaries on the sign, definitions or glosses on the sign in its relation to the referent object. For example, a drawing of a deer, a photograph, or a graph of the birthrate plotted against population size all are interpretants that our ecologist might provide. The production of interpretants is called semiosis. Semiosis is an inherently unlimited process: leading postmodern scholars concluded that no sign, no text, can ever be meaningful and exhaust its own meaning possibilities.

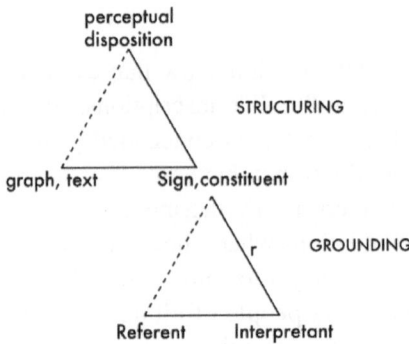

Figure 1.8. There are two stages to the reading of an inscription. During the structuring process, pertinent signs and relationships within a text are identified. During the grounding process, signs and familiar things in the world (referents) are mutually stabilized.

The relation between signs and referents are arbitrary. Even the visual si-militude between the outline drawing of a horse and some living horse is pro-duced, consistent with some cultural decision, and therefore has to be learned. Linking a sign to a phenomenon (or vice versa, a phenomenon to a sign) there-fore requires transformation. These transformations, rather than being *natural* (logic) correspondences, require knowledge of the rules and conventions, r, by which an expression S relates to some content R. Finally, signs never stand on their own but have to be read in the context of other signs surrounding them. All other signs that provide the background to a particular reading of a specific S constitute this constituent c

Reading inscriptions as semiosis

The upper left-hand side of the model represents the reading activity in which the individual attempts to structure the text; that is, the work of reading is inter-nal to the graph. Here, the text (graph, painting, verbal text) itself is the referent, which is structured on the basis of existing attentional and perceptual resources. In this process, the constitutive elements and internal structure of the text are worked out. For example, a reader of figure 1.2.a may identify as signifying fea-tures: maximum of birthrate, intersections of birthrate and death rate, (changing) slopes, (changing) relation of slopes, intercepts with axes, (changing) absolute heights, (changing) relative heights, maximum difference between birthrate and death rate, and so on. Chains of interpretants such as 'birthrate equal death rate' and 'the population doesn't change' (figure 1.2.a) may then be used to elaborate the results of the earlier process. These new interpretants may themselves become signs at the next stage. The nature of salient features identified during reading differs not only between but also within individuals.

Once a graphical feature is perceptually isolated, it can be read as a sign that points to some phenomenon external to itself.[22] The process of connecting sign and referent in reading is called *grounding*. That is, the reader may read a sign in figure 1.2 as a particular statement about a wolf pack in some national park, or about the elk population in Banff National Park. Let us return to several previously used examples. First, Kelly read the minor peak as water associated with a particular rainfall and coming from the northern tributary of the creek. Then, eighth-grade students read the maximum ('optimum') in their graph as the point where negative effects of light on plants begin. Finally, the physics professor (incorrectly) read the level acceleration graph as signifying a constant yearly income. In both stages of the reading process interpretants are understood as the verbal elaborations that each individual provides in the process of engaging the relevant reading. Interpreting a sign therefore means to define a portion of the world to serve as a sign vehicle to denote other portions of the world derived from some global segmentation. In both parts of the model, there is a dialectic process at work, which stabilizes phenomenon and sign, and between an isolated feature and the process of reading. When scientists experience problems with particular readings they may question both levels of reading. For example, one scientist in our database had inferred that a population tended to move to the stable equilibrium. However, in his experience with real organisms, there are always oscillations in population size. He therefore began to question his own perceptual parsing of the initial graph and searched for a feature that would lead to oscillations in the model that he had inferred. That is, he attempted to restructure his perception to come up with a new feature that would lead to an oscillation in his own model.

The semiotic model has two additional features in the sociocultural conventions (r) associated with reading any given sign and the constituents (c) in the graph and associated text (in caption and main text) that both constrain what and how some sign S points to something beyond and other than itself. For example, once a reader has isolated the labels 'birthrate' and 'death rate' on the graph (figure 1.2.a), there are the conventions of the English language how the terms can be used, and therefore what the associated lines can signify. Similarly, the intersections between the two graphs can be read as signs ('birthrate equal death rate') that have to be read in the context given by 'birthrate' and 'death rate' leading to particular inferences. The interpretation of the same intersections would change if the lines were labeled 'birthrate of deer' and 'birthrate of moose'.

Reading an inscription transparently therefore means that the individual is familiar with signs, phenomena, the transformations that link particular signs to phenomena, and the culturally specific conventions constraining the use of signs. Furthermore, the individual has to perceptually structure a graphical display into signs such that they can refer to something beyond themselves. Jon Barwise noted that the problem of the reader is one of finding solutions to a relation with four parameters {S, I, r, c} that determine the content of a sign.[23] Our research shows that even among scientists there is variation in how they perceptually structure a graphical display, and therefore what they read into the graph. Furthermore, the processes are interconnected so that the structuring of a graph is influenced by the reader's familiarity with biological situations and vice versa.

Toward a developmental phenomenology of reading

Reading requires work in order to (a) interpret a text by means of a complex inferential process; (b) check whether or not an expression refers to the actual properties of the things the text is speaking of; and (c) interpret expressions on the basis of certain coded or un-coded circumstances. Texts therefore do not speak for themselves, for they depend on the reader's familiarity with the content domain and cultural conventions regulating the signs that make up the text. There is then a mutually constitutive relationship between texts and the world they refer to, a relationship that is further regulated by the context of the text and sociocultural and cultural historical conventions. This results in constraints on the possibility for development: understanding of phenomena requires understanding of the expressive domain (discourse), but understanding an inscription also requires understanding the phenomena.

Rather than framing the situation as a catch-22 from which there is no escape, development can be thought to occur in the form of a 'bootstrapping' process. Such a bootstrapping process occurs, for example, when members of computer clubs achieve expertise that by far exceeds the initial expertise of any individual member. This bootstrapping is achieved as individuals engage in the activities that provide an experiential basis and the inscriptions (signs) to capture the experience. By engaging in activities that provide students with experiences in the content domain, expressive domain, and translations between the two— conventions being built either by the classroom community or being embodied in the computer technology used—students can then develop their competencies relative to the inscription of the experiences. Students in school similarly develop considerable expertise when provided extensive opportunities in mapping from their own experience and experiments to graphs, and from graphs to their own experience. Thus, with the assistance of teachers or computers, the individuals and classes in these studies bootstrapped themselves into considerable competence.

This model has considerable power, providing viable explanations for different levels of competence in reading graphs, and for the transitions between these levels. Before engaging in a discussion of a more general concept of literacy, let us make another recursion through students' problematic readings of graphs through the lens of the semiotic framework.

Another look at problematic readings of graphs

In our extensive review of the research literature on the topic of inscription we noted that graphicacy takes a marginal place in most commonly used commercial textbooks for the elementary grades. Activities concerning the notational work related to inscriptions cover a few pages of textbook per year. Furthermore, as we show in the next chapter, Cartesian graphs in high school biology textbooks are more difficult to read (interpret) than those in ecology research journals. In the light of this situation, it comes as little surprise that students have difficulties reading graphs.

On the basis of the semiotic model proposed one can predict that if students have had little opportunity to read graphs they should experience difficulties structuring them into signs and contexts, and relating signs into situations (with which they are unfamiliar). Students will not only be unfamiliar with the translation required but also with the conventions (with which scientists are so familiar that they no longer require attentional resources) of sign use. To use an analogy, students are in a situation of someone unfamiliar with German or French asked to read a text in one of these languages. They may begin by spelling out the words, make inferences about the nature of some sign vehicles (e.g. 'Universität', 'université') without, however, being in a position to know what the text is about. In the early stages of reading graphs we would then expect a process similar to early reading: reconstructing words through spelling, and reconstructing sentences in a word by word fashion, without nevertheless understanding what the text is about. One would also expect students to read signs literally: they therefore draw on iconic similarities, because these are the first features we become aware of, rather than on more complex metaphoric relationships between sign and referent, utilize what linguists classify as *faux amis* (false friends, deceptive cognates), and do not 'get' the jokes.

   Earlier we have seen that there is a dialectic relation between a sign that results from structuring a graph and the phenomena it indexes. Students are therefore experiencing more difficulties because they lack familiarity with the referents possible (R). For example, the children in one research project were asked
to interpret a graph in which oxygen levels and shrimp numbers were plotted as
a function of different positions (among which there was a sewage plant) along a
river.[24] Given students' unfamiliarity with shrimp, dissolved oxygen, the relationship between oxygen levels and shrimp numbers, and with the translation
between representations of these situations and the situations themselves, we are
not surprised that many students constructed 'incorrect stories'. This was also
confirmed in one of our studies involving college students. These students perceptually isolated signs, even related some of these to their everyday experience
('death rate', 'birthrate') but remained stuck in literal readings of the tasks
(graph, caption) because of their unfamiliarity with possible content domains
and translations between signs and referents. One might therefore expect that a
person assumes the difference to be perceptually available when asked to identify the larger of two quantities. One might also expect the same to happen if
scientists unfamiliar with mapping a particular graph onto a vernacular situation
may commit errors, especially in situations where they do not have the leisure to
do a careful structural analysis of the referent situation (e.g. in lecture, think
aloud session).
   In fact, tasks that ask students to identify the higher speed from a graph
primarily intended to display distance traveled or distance from some reference
point violate recommendations of good graph design. First, recommendations
for the construction of good graphs suggest that if speed comparisons are the intent, a graph should display speed. That is, respondents are really asked a question that is more easily and appropriately answered in the context of a transformed graph. Furthermore, speed can be read from the distance-time graph
(figure 1.5.a) because the definition of speed and the definition of slope coincide.
The slope of the distance time graph at any point also corresponds to the speed
of the object at that point. If distance were to be displayed as a function of any
other variable, the slope of the resulting curve would no longer indicate velocity.
That is, two conventions coincide. In the absence of familiarity with these conventions, respondents are unlikely to respond appropriately. (The same holds of
course for our own test question to identify from a graph [figure 1.2.a] that
population associated with the largest increase in population numbers.)
   As pointed out above, the sign-referent relation is arbitrary even in the case
of iconic similarity (horse and sketch of horse) and has to be learned. It therefore
comes as no surprise that researchers identified errors commonly referred to as

'iconic confusions'. Students committing these errors take some feature in one domain (sign, referent) and, based on visual similarity, postulate the same visual feature in the other (referent, sign). Thus, given an undulating speed time curve, students often respond by incorrectly attributing each turn in the graph to a curve in the racetrack. In terms of our model, students are unfamiliar with the mapping between the speed of a car on a racetrack and the corresponding graph, they are unfamiliar with graphing conventions, and they do not consider the different contexts of different types of turns in the graphs. When graphs and referent domains are unfamiliar, even highly trained and successful scientists with many years experience read unfamiliar graphs inappropriately, committing types of errors that in the past have been attributed to 'misconceptions' and 'cognitive deficiencies'. Shifting to the model proposed here changes our understanding of the nature of such problematic readings and allows us to construct different foci for the development of teaching strategies to deal with them.

Toward an integrative concept of graphicacy

In the previous sections, we provide an argument for reconsidering graphs as texts and sign complexes, and reading graphs as an aspect of reading inscriptions more generally. The approach highlights the mutually constitutive nature of signs and referents, that is, of knowledge and the world of experience. Knowing therefore means more than acquiring words and texts. Knowing requires the existence of meaningful referent domains that reify the inscriptions; it requires experiences of translating back and forth between experience and inscriptions; and requires familiarity with sign conventions. This view, therefore, has considerable implications for classrooms. Simply exposing students to new (forms of) inscriptions through books and lectures will not do.

The semiotic and critical hermeneutic approach to reading taken here allows us to move towards a more integrated concept of literacy concerned with inscription practices more generally. The semiotic model is certainly not only relevant in the context of line graphs or other forms of mathematical inscriptions but also concerns other forms of non-verbal inscriptions such as maps, graphical models of processes, and photographs. Such inscriptions are used in textbooks across subject areas. Greater attention to and research about how students read and interpret these variegated multi-modal texts is certainly necessary. Such attention is particularly necessary with the proliferation of visual imagery in the media (e.g. TV or internet) and therefore in students' everyday lives. Integrating inscriptions and other non-verbal signs (e.g. gestures) into a comprehensive semiotic theory leads to a new and different concept of literacy.

The previous considerations therefore lead us to a concept of literacy as the competence to engage in the work of reading any text, by identifying signs, possible referents, audiences, conventions of sign use, and other ways of structuring texts. In this way, (scientifically, mathematically, or artistically) literate students are empowered to engage in semiotic activities more generally to the point of interrogating the presuppositions underlying the forms of inscription chosen by the author(s) of a given text. Educators sometimes are led to despair because students do not have the same cultural backgrounds. However, all students have already developed both experiences and representative forms to deal with these experiences. If we view actions and culture as texts then all students, even if they live in housing projects and inner-city districts, have an experiential basis from which literacy can develop. That is, rather than beginning the learning process with texts that typically describe white middle-class culture, school activities might well begin in the street. Students fail academic science even though their teachers try to connect the subject to students' experiences and simultaneously experience success while producing science in the street. Rather than beginning with the academic subject matter and linking this to vernacular experiences, educators may want to start with vernacular experience and then move, through reflective and critical readings of this experience, to a more academic understanding of it.

This concept of literacy brings to the foreground literary traditions, conventions, and engagement in activities that mediate linguistic and sociocultural practices and decreases the focus on individual knowledge deficits. 'If we view the world as a text, then literacy means engaging the full range of what is in the library (conventional notions of reading), the art gallery (the making and interpretation of art), and the street (popular culture and student experience)'.[25] One primary objective of education would be to allow students to appropriate the (semiotic and hermeneutic) means for constructing and deconstructing literary, artistic, and cultural texts and to acknowledge that meaning is not fixed but changes as texts are structured in different ways. Students are then empowered to undertake a dialogue with others who speak from different locations (centre, margin), histories, and experiences.

The semiotic model allows us to theorize learning and development across the curriculum in a more integrated manner. Sociocultural models have postulated connections between linguistic knowledge and sociocultural (worldly) knowledge. Most important to the development of graph-related competencies, sociocultural knowledge (about worldly objects and events) and linguistic knowledge (about sign forms) are mediated in and through the activities an individual engages in. Thus, 'sociocultural and linguistic knowledge structures activity, and activity creates . . . and recreates . . . knowledge in both of these domains'.[26] If students are to develop both semiotic knowledge and knowledge about those objects that texts are said to be about, teachers need to make available suitable activity structures that allow development and growth. One might then expect—similar to linguistic development in children where signs (e.g. words) initially linked to objects gain increasing independence—that students will develop graphing competencies that become independent of particular situations in which they have worked.

Critical graphicacy and border pedagogy

Conceptualizing verbal texts, graphs, drawings, photos, and other communicative devices as texts provides the basis for a notion of critical graphicacy that applies across the curriculum. We would therefore expect students to be empowered if they engaged not only in learning subject matter of history, language, science, and mathematics but also in (semiotic) analyses of different forms of representing. Students no longer learn only the 'right stuff', but also learn how sign forms (inscriptions) support particular readings, argument, and politics. They also learn that changing sign forms will carve up the world in different ways and, therefore, that meaning is neither transcendental nor fixed or terminable. Critical graphicacy thus defined becomes epistemological and discursive praxis; literacy becomes part of a broader project of border pedagogy and a reconstruction of democratic public life.

The framework outlined here changes the way educational researchers analyze data and identify cause and effect relationships. Regarding the research on graphing, semiotically informed researchers focus their attention on students' familiarity with (a) sign forms, situations in the world, and the translations between the two, (b) disciplinary conventions regulating sign use and possible interdisciplinary differences, and (c) perceptual analysis of sign complexes. Because of the arbitrariness of sign-referent relations and the conventional nature of syntactic and semantic properties of sign forms, the sense of inscriptions cannot be inferred on the basis of their properties alone. The semiotic model then decenters research from making attributions to individual minds to focusing on students' experiences with inscriptions.

The semiotic model has also implications for the way we conceive of the production of texts. Thus, a semiotic perspective alerts writers (students, teachers, educational researchers) to attend to the readers' processes of interpretation rather than assuming that graphs are inherently well or poorly constructed. Producers of texts need to think about the reader and the politics of the texts they release. Texts (graphs, paintings, web pages) do not speak for themselves but are read given the articulated and non-articulated constituents of those graphical sign elements important to the writer's desired interpretation, the conventions regulating sign use within the reader's community, and the familiarity of the reader with the referential situation. Once students understand this relation between inscriptions and reader, they can become more critical with respect to their own use of graphical representations, and thereby enact *critical* graphicacy.

2   The work of reading graphs

Practices related to graphing are quintessential to scientific endeavors of inscription. Graphs are extremely useful to scientists because they (a) constitute the best tools to articulate covariation between continuous measures and (b) are useful to summarize large amounts of data in economical ways. Our own research showed that graphing is one of the key skills of professional biologists.[1] Thus, especially in light of current reform movements that focus on 'authentic school science', it appears important to organize school curricula so that students participate and develop competence in graph-related representation practices of science as the basis for developing epistemological competence involved in *critical* graphicacy. We begin by specifying a brief ontology of graphs, which articulates the kind of things that are important during interpretation.

A brief ontology of graphs

Ontologies specify (a) the things and their aspects that make up the universe under study and (b) the relationships between these things. In our account, we focus on those aspects of Cartesian graphs that are less frequently or never considered in research on graphing.[2] This ontology will allow us to understand some of the difficulties students have been shown to have, but which could not be explained by traditional research other than referring to cognitive deficiencies.

Aspects of Cartesian graphs

Line. The epitome of a Cartesian graph is the line graph; not so much the entire graph but the line used to express the relationship between two sets of measurements. In situations where interlocutors share or assume to share the same situation definition, line graphs frequently reduce to traces gestured in the air (figure 2.1.h). Because of their interpretive flexibility, the meaning of such traces is irremediably situated and contingent. A more elaborate form of the trace is one made on a chalkboard or whiteboard including a set of orthogonal axes (figure

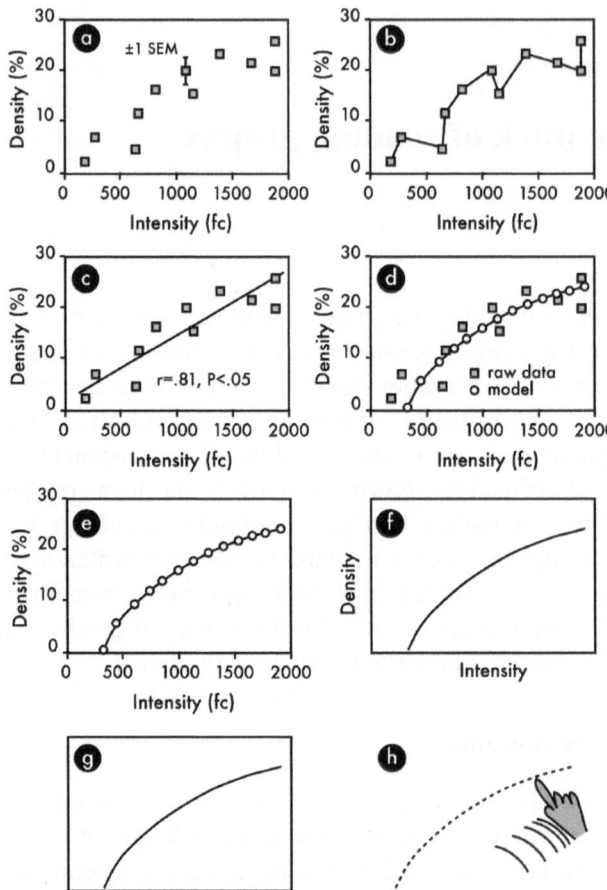

Figure 2.1. Categories of graphs developed during this study included (a) scatter plots including data points only; (b) scatter plots with line connecting points; (c) scatter plots with best-fit curve (and/or statistics); (d) scatter plot with plot of mathematical model; (e) graph of mathematical model; and (f) graphical model without scales or units. Two separate categories that had been identified in previous research were included to complete our category system. (g) Curves accompanied by axes only and (h) curves rendered by means of gestures during face-to-face conversations.

2.1.g). Here, the trace has been situated in a rectangular frame of axes that span a space of yet-to-be determined nature. Because of the conventions in Western scientific culture, we can take this graph (figure 2.1.g) as the root image for inscriptions that appear in publications. The inscription still lacks resources that allow a reader to situate it in a particular scientific context, and to re-construct the space that the author created. The labels on the axes of figure 2.1.f have created this space. Thus, points further to the right generally mean 'higher intensity', whereas points further up generally mean 'higher density', and vice versa. Through additional features, this root image receives further specification (figures 2.1.d, 2.1.e).

Background. The background of a Cartesian line graph is an aspect to which few researchers attend. But it is the lack of detail in the background that makes for the power of graphs. In graphs, the background that still appears in many other inscriptions has been reduced to the whiteness and cleanness of paper; there is no more the 'gratuitous detail' of photographs and drawings that lead the reader astray from the goals of the text.[3] Photographs and drawings offer more opportunities for alternative interpretations that scientists have to bar by appropriate arguments in the text.[4]

Whereas the background in photographs and naturalistic drawings and 'naturalistic' details in diagrams provide illusions of a continuity of the representation with our everyday lived experience (see chapters 4 to 6), this illusion no longer exists in graphs. Our ongoing research among scientists shows that their level of competence in interpreting graphs is related to the degree with which they are familiar with the inscription, the context of data collection, and the translations required between natural setting and inscription.[5] Stripped of 'gratuitous' detail in the background, Cartesian graphs become ambiguous (especially to newcomers) with respect to their referents; that is, they obtain an increased interpretive flexibility as to what they reference and what they express. Readers have to reconstruct the referent and the absent continuity with their lifeworlds from the details of the accessible surface of the display. 'Realistic' phenomena that are continuous with our lived experience have to be recovered from the dot-like existence into which scientists have reduced them. This process of recovering reality from lines and dots lies at the origin of many difficulties students and scientists face when they try to understand the role of graphs in the textual demonstration.

Cartesian graphs, like other inscriptions, use both dimensions of the flat surface of paper. A third dimension is sometimes simulated through conventions that simulate a three-dimensional view. However, this form of inscription brings with it a new set of potential problems. These problems were topicalized in many of the drawings by the Dutch artist M. C. Escher. However, Cartesian graphs exploit these two dimensions differently from other inscriptions such as photographs and naturalistic drawings. Photographs, naturalistic drawings, and

some diagrams share with readers' everyday lifeworlds the nature of spatial relationships. In this way, these inscriptions are 'more realistic' than graphs; these
inscriptions maintain the spatial meanings of the two dimensions of the medium.
Graphs, on the other hand, create new spaces that may have little to do with the
space of our experience. Both axis labels in figures 2.1.a-f, light intensity and
bramble density, are new dimensions projected onto the linear extensions of
space spanned by the two axes. That is, the graphs in figure 2.1 profit from the
exploitation of the two available dimensions. However, through the superposition of arbitrary labels, these dimensions are no longer spatial but can refer to
anything that a scientist deems relevant, for example, light intensity, bramble
density, time, energy, or velocity. Readers can use the axis labels that come with
many graphs for discovering the nature of the created spaces and their dimensions.

Cartesian graphs have particular affordances that other forms of representations such as language do not have. Graphs allow the depiction of continuous co-variations between two sets of measurements (with pairs of measurement taken on the same object). Because graphs have a topological quality
that contrasts the typological quality of words, they are much better suited than
language to register the continuity with which we experience the everyday world.
However, graphs cannot be understood simply by studying the continuity of the
displayed curves, and contrasting their topological quality with those of words.
Graphing practices, like other human practices, mix typological (here labels)
with topological (here lines) representations. Before scientists study a relationship, they must make a decision as to the nature of the variables of interest. The
identification of variables embedded in the axis labels, however, involves
choices of typological nature.

Scales on the axes and *units* associated with the labels further reduce the interpretive flexibility associated with a simple line as the contrast between figures
2.1.e and 2.1.f illustrates. Scales on the axes allow the mapping of measures
from the instrument to the graph. They give the illusion of continuity between
'phenomenon' and graph, but barely cover the ontological gap between the instrument and the graph, let alone the graph and the worldly object that was in
contact with the instrument. The scales re-present the measurement; they define
a metric of the new space, and the link to the instrument. Units function in a parallel way to scales in that they assign orders of magnitude and quality to the
numbers on the axes. An axis labeled bramble density, and scaled in tens from
zero to fifty is open to interpretive flexibility. A value of twenty could mean:
twenty percent coverage, twenty plants per square meters, or twenty percent
near-read reflectance. (Reflectance is a useful index for the canopy-level leaf

area of a species.) Scales and units therefore provide resources for the reader to move from the two-dimensional, imaginary space to actual measurement instruments. They are resources to link, at least for many phenomena, the paper world with the experienced lifeworld.

Scientific processes in the laboratory or field form an integrated web of signification such that they lead to a mathematization of nature. Thus, 'raw' data are plotted onto a set of Cartesian coordinates, only after the decisions about the nature of the variables and instruments have been made. 'Scientific' data collection is, by its very nature, geared toward mathematical representations such as measurements, counts (frequently represented in tables, histograms, and bar charts), equations, and graphs. (Of course, mathematical representations are part of a scientific ideology, and a reason why 'hard' scientists frequently belittle those researchers concerned with arriving at understandings of our world based on hermeneutic and other non-mathematical understandings.) Each point in a plot corresponds to one unit of analysis. Although based on 'real' data, each data point constitutes a new entity in a new, theoretical space created by the axes and labels.

The data point re-presents the unit of analysis, an individual or thing. It is this original individual or thing that readers of the graph have to recover through the act of their reading. A small dotted circle in figures 2.1.a-d stands for two measurements: light intensity and bramble density taken in a different plot of the research site. Each dot, therefore, corresponds to a different piece of land where pairs of measurements were taken. In such cases, the underlying assumption usually is that the unit of analysis really represents a class of objects and each data point is an example from the entire class. All data points in figure 2.1.a therefore stand in for (re-present) objects that are members of the class (or universe); the relationship, which obtains for the class, therefore obtains for all its members. However, different data points do not have to re-present classes of objects, but can stand for one individual or thing. For example, if the length of the same leaf was measured over a period of weeks or months, these measurements could then be plotted against a horizontal axis representing time. The points would then stand for 'the same leaf' but at different points in time.

Best-fit procedures are based on the belief that it is possible to recover the continuity of some phenomenon from a small number of measurement pairs. When relationships and laws are derived from data points by means of best-fit procedures (as in figure 2.1.c), continuity is constructed rather than directly accessed. (Continuous graphs can be recorded with devices that 'directly', without analog-to-digital conversion, translate between phenomenal world and paper; examples are thermistor-based temperature or motion detectors used in microcomputer-based laboratory equipment.) The continuity of the best-fit graphs has been constructed by making disappear the discrete and categorical practices of naming and assigning, choosing a unit of analysis, and collecting data points. That is, scientific practice covers up the ontological gaps existing between consecutive representations on their trajectory from nature to mathematical law. Some parts of readers' interpretive efforts are devoted to the recovery of natural situations (phenomena) for which the graphs are reasonable renderings.

Cartesian graphs frequently include (a) legends (figures 2.1.d and 2.3 below), (b) points of different color, shape, and size (figure 2.1.d), (c) lines of different nature (solid, broken, dotted), thickness, or color (figures 2.3 and 2.6 below), and (d) lines marked with arrows and labels (figures 2.2, 2.3, and 2.6 below). All of these aspects of Cartesian graphs are used as *markers* that allow authors and readers to coordinate their orientation in reading and interpreting. In contrast to the topological aspect of a line that signifies the relationship, these markers are of a typological nature. These markers allow authors to foreground one particular line or a set of data points with a minimum of verbal description. They can also be used to instruct readers to read graphs contrastingly, that is, to compare different aspects present in the same panel.

Graphing resources

The aspects of Cartesian graphs just described (e.g., axis labels, scales, units, and so on) can also be understood in another way: They constitute resources for making sense of the primary curve that traces the functional relationship between two sets of corresponding measurements. They also constitute informational resources and therefore reduce the interpretive flexibility of the primary line or series of data points. To tease patterns from the data they collected, authors have available a wide range of resources. These resources are particularly important when the inscriptions serve demonstrative purposes in scholarly texts and textbooks. Some of these resources include: axis labels (figure 2.1.a-f), nature of scales which can be log-log, linear-log, log-linear, linear-linear (figures 2.1.a-e), units ('foot candles' in figures 2.1.a-e), legends, error bars (figure 2.1.a) or confidence ellipses, titles, and statistics in figure (figure 2.1.c), caption, or text; color (figure 2.1.d), shape, and size of data points; color and nature of lines (broken, fat, dotted); labels to curves, arrows (figures 2.2, 2.3), and other identifiers such as numbers and words; or juxtaposition of lines within the same graphs and series of graphs (see our figure 2.1).

These resources assist readers in a reconstructive process that allows them to imagine 'real' situations from which these graphs have been abstracted. Scientific authors usually draw on these resources to limit the interpretations of the reader to those they intended:

Scientific illustration behaves exactly like the scientific article itself: it is constructed like some military strategy, an ambush without an escape route. Each time that a reading of the results different from that of the authors could possibly be made, the departure is barred by an adequate argument.[6]

However, even with these resources that limit interpretation, the reconstructive process is inherently open because each graph can have many different meanings. Such interpretive flexibility is unavoidable even in the case of simple diagrams. A diagram of eight lines forming a schematic cube can be read in many different ways: a glass cube, a wire frame, three boards, an open inverted box, and so on. How the diagram should be read depends on the context, in particular the interpretive resources made available by the author. The fewer the resources, the greater the interpretive flexibility, and therefore range of situations from which the graphs could have been derived.

Cartesian graphs, captions and main text

Graphs do not stand alone, but are related to two kinds of text, the caption and the main text. Graphs are frequently treated as adjuncts to the text, as material basis, 'proofs', of the claims. Semiotic analyses treat graphs and text more symmetrically and emphasize the mutually interactive nature of the two types of communication. Cartesian line graphs express what words cannot express or can only express in cumbersome ways, such as continuous variation of two associated sets of measurements. (By means of mathematical expressions of typological nature [e.g., 'parabola' or $y = m \cdot x + b$], continuity can also be expressed but is often less accessible even to mathematicians and scientists who increasingly use computer-based visualization tools.) Text usually handles more easily the construction of distinctions. Here we are concerned with the interplay of graphs with additional texts in the form of captions and the main text.

Captions provide additional resources to the reading of graphs, and thereby further limit the interpretive flexibility of graphs. A sample caption was constructed for one of the graphs in figure 2.1 (see figure 2.2). A cursory reading provides the following: The first sentence of the caption describes the contents of the graph as a functional relationship between bramble density and light intensity in various plots. The following two sentences provide readings of specific parts of the plot, in this case, its two extremities. Finally, the last sentence indicates the source of the second set of data points and curve.

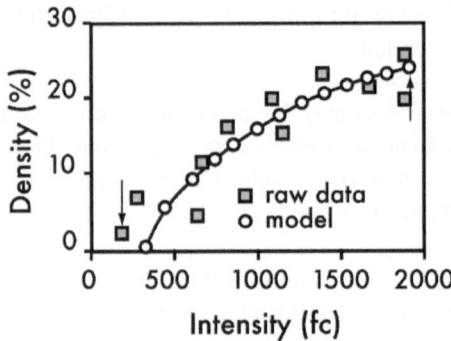

Figure 1. Bramble densities in various plots at one riparian site as a function of
light intensity (squares). Below the threshold of about 250 foot candles, bramble densities level off. Plotted fro comparison are predicted values from the optimum absorbance model (circles, line).

*Figure 2.2. Illustrative example of a graph and caption. Arrows and the text in the cap-
tion provide both readings and instructions for how to read the graph in a specific way.*

However, this caption does more than state the obvious. First, it provides
context by indicating the source of the data from 'one riparian site'. Furthermore,
it suggests that the unit of analysis is 'various plots'. Second, the middle two
sentences guide readers' attention to particularly relevant places in the graph, the
nature of the observed relationship at the two extremities of the plot. That is, the
caption provides one reading, a description of the graph's content, namely that
favored by the author. However, captions do additional work: they serve as instructions for how to read graphs, particularly the graph with which they are associated. For example, that graphs express relationships between the measurements indicated on horizontal and vertical axes can be taken as a shared
understanding among scientists. However, for someone unfamiliar with graphing, the statement is an instruction where to direct attention and how to read.
The arrows that direct (in an instructive sense) readers' attention to those points
mentioned in the caption text highlight the instructive nature of the middle two
sentences: attend to key points in the graph. Main texts provide additional descriptions of, or instructions for, reading the graphs.

Main texts provide the story line in which graphs with their captions are
embedded. They provide the story lines that set up a frame for graphs and their
captions. Main texts may supply readings of the graph, and therefore instructions for how particular graphs can and should be read; they put graphs or as-

pects of graphs into relief and therefore set readers up for the interplay between the multimodal nature of concepts. Finally, main texts provide interpretations that integrate the graph into a theoretical framework. If graphs are not referenced from within the main text, they become ancillary to the text's argument, and there may be little reason for readers to attend to this additional material.

However, there is a difference between the two forms of text associated with graphs. Whereas captions are always located next to graphs, the spot in the main text where authors refer to their graphs is most frequently dissociated with the location of the graph/caption. In some instances, reference to a graph and the graph do not appear on the same page. Nevertheless, main texts draw on graphs as the material bases for the claims authors make; graphs form rhetorical resources in the demonstrative practices of science. Because graphs can be used to express continuous variations, they are more economical ways of reporting relationships than text alone. This supportive function of graphs in the main story is, however, only part of the relationship between graphs and text. Main texts, similar to captions, limit the interpretive flexibility associated with graphs that stand on their own. Main texts highlight and therefore bring within readers' horizons those aspects of the graphs that authors deem necessary within their argument. Main texts can therefore be considered as providing authors' readings and interpretations. But main texts, like captions, have another side: they constitute instructions for reading and interpreting graphs.

A reflexive note is appropriate here. Our figure 2.1 is on a page different than the text that refers to it. Both figure and text exemplify the categories and constructs we use. The text talks about the categories; figure 2.1 authenticates and exemplifies them. However, in contrast to the popular adage that a figure is worth ten thousand words, we suggest that the text and figure need one another; they feed on one another, each giving authority and authenticity to the other.

Although main text and caption provide constraints, they are still open to alternative interpretations of the instructions for reading, and as descriptions of how to read the graph: 'The description of something real is always imprecise; there is always a gap between a description of reading and reading's work'.[7] The gap between description of the graph and reading the graph has to be filled by the act of reading itself. It is through the act of reading that readers establish the congruence between the two organizational classes of materials, text and graph. Scientific authors construct their main text and captions to provide particular constraints to any reading of the graph. Main text and captions are constructed such that they point to those features that the author needs in the bricolage of their demonstration. Main text and caption therefore work in the same way as the resources within Cartesian graphs: they are designed to bar readings of the results different from that of the authors. The text will figure as a specification of order that the reader seeks and recovers through competent reading of text and graph. In this, text and graph ascertain the visibility of the claimed natural object. To achieve this, text and graph have to allow readers to construct them in congruent fashion. Finally, main text and caption do not only provide descriptions (directions to read), but also receive evidentiary support from the graph. The relationship between graph and (caption and main) text is therefore reflexive. Being from different organizational classes, graph and text constrain and draw support from each other.

The ontology of Cartesian graphs, captions, and their associated main text developed here provided us with a resource to our anthropological approach of analyzing the forms and functions of graphs in textual pedagogy. The following four case studies illustrate differences between Cartesian graphs that appear in ecology journal articles (case study 1) and those that are used in high school biology texts at different academic levels (case studies 2–4).

Graphs in ecology-related journals

For this investigation, five journals were selected because they represented a broad range of foci: applied research, pure research, and modeling ecological processes: *Journal of Animal Ecology, American Naturalist, Ecological Monographs, Ecological Applications*, and *Ecological Entomology*. Both formal (citation indices) and informal (biology professors) indicators suggested to us that our selection represented highly regarded journals. Three of the four journals categorized under the science citation index entry of ecology ranked among the top ten journals; these ratings were consistent for the seven-year span we investigated (1988–1994). Ecological Entomology ranked seventh in 1994 among sixty-two entomology journals in the science citation index, a position it consistently held over the same seven-year span. For each journal, we sampled the entire 1995 volume if the total number of pages was less than 500, or the number of articles (beginning with the first article of 1995) to bring us above the first 500 pages of the volume.

General observations

In the scientific articles we surveyed, graphs (like other inscriptions) are well integrated with the text. These graphs feature scatter plots of data, scatter plots of data with lines connecting points, scatter plots with best-fit curves and statistics, scatter plots and results of mathematical models, or output of mathematical models only (figures 2.1.a-e). Graphical models that do not include scales and

units (figure 2.1.f) are virtually nonexistent in the journals surveyed. Graphs seldom are presented alone, but in figures that group two to six graphs together,
much like our figure 2.1. The joint appearance of several plates of graphs, or
several line graphs within the same plate allows authors and readers to make
contrasts. It is important to note that the contrasts are not made by the figure, but
in readers' interpretive engagements with the text and the plates.

Captions are frequently extensive—sometimes running from 150 to 200
words. The caption directs readers to specific features in graphs that are relevant
in the present argument of the authors. Captions are descriptions of the graph
and, when mathematical models are also featured, frequently provide statistical
information about the goodness of fit between data and model. Scientific articles
provide contextual information and 'instructions' on how to read the graphs so
that these interact with the text to constrain meanings. Authors do not assume
their graphs to provide visual proof for a phenomenon, but do everything to assure that the graphs they supply are read in unambiguous ways (see the quote in
the previous section). On the one hand, some may argue that this information is
redundant. Whatever the authors provide in terms of information is already
available in the graph. On the other hand, these 'descriptions' can also be seen
as instructions for how to read the sequence, which numbers to focus on, and
which aspects of the graph to attend to. The following analysis points out central
elements of graphs in the scientific literature: Authors provide many resources
that reduce interpretive flexibility and facilitate readers' construction of meaning
such that they find the argument in the main text supported by (and congruent
with) the associated graphs; furthermore, main text, graphs, and captions are integrated and provide sufficient redundancy to assist readers in the construction
of convergent meaning between the different parts of a research article. That is,
readers of scientific journals find a periphery that contextualizes graphs and
thereby allows specific readings to emerge; that is, readers of scientific journals
find that genre borders provide sturdy yet light scaffolding for the ongoing
graph-related activity.

Case 1: Hypothetical survivorship curves in ecology journal

Our sample graph and main text (figure 2.3) from the scientific literature comes
from an article on livestock-mediated tree recruitment in Kenya. In this article,
the authors present a number of investigations to show the impact of herding on
the generation and regeneration of woodlands. Compared to other articles in the
journal, the authors employed a relatively high number of scatter plots with connected data points.

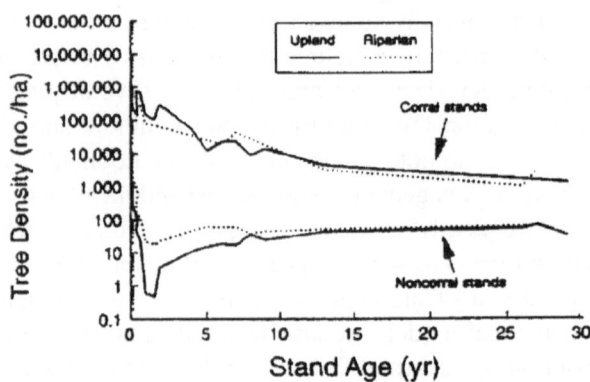

FIG. 9. Hypothetical survivorship curves for corral and
noncorral trees created from tree density data (no./ha, log
scale) in stands of different ages found in riparian and upland
areas.

*Figure 2.3. Graph with caption from a scientific journal article. (Reprinted with permis-
sion of the copyright holder.)*

On first glance, readers may note the large number of resources that accompany the four line graphs. Of the four lines, two are dotted and two solid, and
they appear in pairs of a dotted and a solid line each. The legend makes the distinction between solid and dotted lines. Additional markers ('corral stands',
'noncorral stands') distinguish the two coordinated pairs of lines. The horizontal
axis is marked off in increments of five, which the label instructs us to read as
'stand age' measured in years. The vertical axis immediately draws attention to
the scale that is equally spaced for consecutive orders of magnitude: It is a logarithmic scale. The label instructs us to read the scale as 'tree density' in units of
number per hectare (no./ha).

The resources provided in the graph itself are, for experienced scientists, already enough to construct their own interpretations. The markers ('corral' and
'noncorral') and legend ('upland' and 'riparian'), along with the different types
of lines, suggest a two (type of stand) by two (distance from river) design. Such
a design suggests to the scientific reader to make comparisons that would reveal
main effects and interactions.

This *caption* illustrates that scientists do not leave uncertain how to read a
line graph. It repeats information that might be gleaned from the graph alone.
The effect of this redundancy, however, is to guide readers to a congruent con-

struction of graph and text. The marker 'hypothetical' sets this graph apart from all the others that the authors had used. The caption explains how this hypothetical curve of survivorship was constructed: 'created from tree density data in stands of different ages'. Here, the unit of analysis represented by each point on the line is identified: different stands of trees. The caption alerts readers to attend to the scale of the logarithmic tree density data, although a look at the ordinate may have revealed that the distances between 1 and 10, 10 and 100, 100 and 1000, and so on are equal, from which we can infer the logarithmic scale.

In their *main text*, the authors provide a reading of (or instructions for reading) the graphs in their figure.

The density of older trees in corral and noncorral stands representing many different cohorts can be concatenated to create hypothetical survivorship curves for A. tortilis (Fig. 9). The curve for the corral stands accurately portrays the survivorship of the corral cohorts that we sampled. The curve for the noncorral stands is only generally representative of survivorship of these trees, because, unlike corral trees, we were not able to estimate the age of trees in noncorral stands with complete accuracy once trees exceeded 2 yr of age (see Methods). Inaccurate age estimation left to the obvious artifact in the survivorship curve for noncorral trees, showing an increase in density after 2 yr. Despite this shortcoming, the curves do display some interesting generalities. The overall shape of both curves is as expected: mortality is very high in young trees and low in older trees. But the slopes are different; noncorral stands initially appear to thin much more rapidly than corral stands, due to high mortality in the first 2 yr, whereas corral populations fell by only one order of magnitude.

Mortality patterns in older trees, however, appear to switch. In corral stands, older trees appear to die at a fairly constant rate, even when trees are >10 yr old. This would be expected if corral stands, because of crowding, thin consistently over time. By contrast, older trees in noncorral stands show very little mortality after 10 yr, if any. This suggests that mortality is indeed somewhat higher in the crowded corral stands when trees are beyond 2–3 yr of age.[8]

The authors explicitly identify the unit of analysis plotted in the graph, different stands, representing different cohorts concatenated to give hypothetical survivorship curves. They then describe the reliability of the data displayed. For corral stands, their age estimation is fairly accurate, so the data are 'accurate'. For the noncorral stands, age estimation was less certain. The text refers readers to consult the methods section for a description of the estimation procedures. The impact of inaccurate estimation procedures on the graphs is immediately described: an increase in tree density after two years is visible from the graph. This is also an instruction for reading the unexpected increase in tree density at about two to three years in the graph of noncorral stands. The text explicitly marks this artifact as a shortcoming.

The main text then provides a description (and thereby instructions for reading) the graph. The text begins with a description of overall shape of the curves in terms of mortality (decrease in density), and then directs readers' attention to the differences between the curves for corral and noncorral stands. More specifically, it directs readers to attend to the different slopes of the curves during the first two years, and explicitly marks the decrease for corral populations as 'one order of magnitude'. Because of the semi-log representation, one order of magnitude corresponds to one major tick, and is therefore easily accessible as a fact.

In the final section, the text provides a comparison for older trees. Again, a contrast is offered between corral and noncorral stands: 'Mortality patterns . . . appear to switch'. This statement is subsequently elaborated: 'In corral stands, older trees appear to die at a fairly constant rate'. 'Die' is an instruction for assessing the line graph in terms of a decrease in density, and 'constant' instructs us to attend to the linear tail of the line for ages greater than ten years. The text then offers the earlier announced contrast: 'Trees in noncorral stands show very little mortality'. Here, we are directed to attend to the constant density of the trees, associated with a horizontal line. An overall reading and interpretation is then offered for the two different trends: When trees are older than two to three years, the mortality rate is larger in 'crowded' corral stands.

There are other indicators in the text that suggest to readers the tentativeness of scientific knowledge rather than the presentation of ready-made facts. For example, the phrasing 'the overall shape . . . is as expected' signals a comparison with an earlier expounded hypothesis according to which high densities lead to larger rates of thinning (and mortality); 'appears' highlights uncertainty; 'this suggests' marks the contrast with the earlier description of the graph. Here, inference and description of the graph are clearly separated.

In our reading of the ecology journals, this example is rather typical in the tight integration of graph, caption, and main text. The integration is tight, in part, because from an informational point of view, the three parts of a scientific article are redundant. This redundancy decreases the interpretive flexibility of any single aspect and assures a greater likelihood of intersubjectivity between authors and their readers. Our analysis shows that scientific authors construct this graph-caption-text triad such that it permits but one reading. In this, our analysis confirms the characterization of scientific illustrations cited earlier: Each scientific illustration constitutes an informational ambush that leaves no escape for alternative interpretations. Rather than letting readers interpret the graph, the caption directs them to particular aspects: the unit of analysis, the hypothetical nature of the survivorship graph, the different sampling sites (riparian, upland, corral and noncorral), the different ages of the tree stands. These particular aspects are also available as resources in the frame of the graph itself. The main text provides a particular reading and therefore instructions about how to read the graph. Thus, although the audience of this article is definitively scientific, descriptions of (instructions for reading) graphs are provided, and attention is called to units and scale. This is a practice that, as we show below, is not the rule in high school textbooks or in research on students' competencies related to graphing.

Graphs in high school biology texts

For the present investigations, we included six, in North America widely used biology textbooks. Three of these textbooks were among the four selected for their representative nature in a study of textbook structure.[9] Our selection also includes a range of target audiences including a general biology text for students with low reading levels, texts for introductory biology students, and texts for more advanced biology students. In all textbooks, we identified those chapters that related to ecology for the comparative analysis with scientific journals. However, we included the entire textbooks for the conceptual analysis of Cartesian graphs.

General observations

Our analysis of the ecology sections in six high school biology textbooks revealed that the relative frequency of plots using Cartesian coordinates is surprisingly low compared to the utilization of the same inscriptions in the corresponding scientific literature. A more detailed analysis of three of these texts revealed that the total number of Cartesian graphs is low. There were only thirty-four, twenty, and four graphs in the entire textbooks, with decreasing frequency for advanced, introductory, and general level texts. There was also a particular predilection for line graphs that represented models, including few graphs that included data from actual experiments; frequently, graphs lacked units and scales. In some instances, graphs were not even referenced to in the main text. Even ecologists have difficulties interpreting such graphs.[10]

Graphs come in many different guises, vary in the number and kind of resources they make available for interpretation, differ in the way they are integrated with caption and text, and are part of different scientific concepts. It is therefore difficult to make generalizations beyond those already made. We chose instead to present three case studies, one each from the three levels of audience: advanced, introductory, and general. These examples also show an increasing specification by the associated texts and therefore provide an indication for the continuum of graph-text featured in textbooks (though frequencies are heavily slanted in favor of those represented by case 2). The first two case studies (case 2 and 3) concern graphs rather typical for high school biology textbooks; the graph from general level biology (case 4) is different and therefore not representative, but constitutes a positive example how to use and integrate graphs into high school texts. (Such cases in qualitative inquiry constitute 'negative cases'.)

Case 2: Natural selection in advanced biology

Our sample graph with associated caption (figure 2.4) and main text comes from the section on natural selection in an advanced biology textbook.[11] The graph with caption and main text appears on the same page. A cursory look at the figure reveals three panels labeled 'a', 'b', and 'c'. The line common to all three panels is colored brown. Panels 23.13b and 23.13c each also feature a second, black line. In Panel 23.13b, the black curve is more peaked than the brown curve, whereas both curves are the same shape in Panel 23.13c but the black curve is shifted.

This figure is typical of many that appear in high school textbooks: it is marked by the absence of many resources that would assist in reading. In this, it is similar to the graph in figure 2.1.g, which, as stated earlier, can be observed in face-to-face communication among scientists. We did not observe one single example of this type of graphs in the scientific journals we surveyed. That is, resources that are made available on a routine basis in the scientific literature have been omitted here: there are no titles, labels, scales, units, or other textual markers. There are no indications what the horizontal axis stands for and the vertical axis is absent altogether. As such, this figure could be used in many other situations and domains, for example, in statistics courses where students learn about distributions, the underlying logic of *t*-tests, and so on. Because the graph lacks specificity relative to a particular domain, measurement, or concept, readers must seek recourse to caption and main text in order to construct meaning for this figure.

The first sentence suggests 'two' types of selection that contrasts the 'three' panels below the caption. The engaged reader may expect a resolution of this conflict to arise from the stable brown curve contrasted by the two black curves in Panels 23.13b and 23.13c. Further tensions arise with the second sentence that instructs us to read Panel 23.13a as 'Initial distribution of the phenotypes present in a population'. The word 'phenotypes' suggests a typology that, as all typolo-

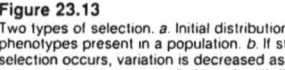

Figure 23.13
Two types of selection. *a.* Initial distribution of the phenotypes present in a population. *b.* If stabilizing selection occurs, variation is decreased as the extreme phenotypes are eliminated. *c.* If directional selection occurs, an extreme phenotype is favored over other phenotypes.

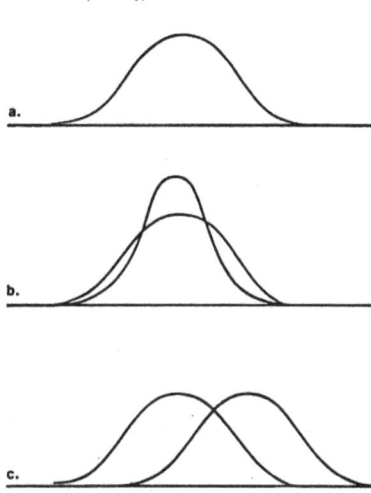

Figure 2.4. Cartesian graph with from a high school textbook for an advanced-level biology course. Graphs and caption lack important resources for understanding the topic of stabilizing and directional selections. (Quoted and reprinted [scanned graph] with permission of the copyright holder.)

gies, is categorical. On the other hand, the line of the graph suggests continuous variation. If the horizontal axis encoded typology, a bar chart would be more appropriate.

Readers find out in the next sentence, referring to panel 23.13b, that 'If stabilizing selection occurs, variation is decreased as the extreme phenotypes are eliminated'. Here, the author does not describe what her panel 23.13b displays, but more generally about selection and decrease in variation. To make sense of this, readers need to be familiar with representation practices in statistics according to which the width of a distribution refers to variability. A sharper peak refers to a narrower distribution and smaller variability. The third sentence instructs readers to understand panel 23.13c as an instance of directional selection: 'If directional selection occurs, an extreme phenotype is favored over other phenotypes'. Again, the caption presents information not directly available in the line graph. Furthermore, the use of 'phenotype' in its singular form suggests one category that is favored. However, the graph featured in the corresponding panel does not display types; rather, it displays continuous variation.

It is evident that the caption does not provide a reading of the graph, or instructions that allow readers to construct direct links between the descriptions of panels 23.13b and 23.13c with the lines themselves. For example, the sentence 'If directional selection occurs, an extreme phenotype is favored over other phenotypes' does not describe what can be seen in panel 23.13c. Rather, possible meanings and relationships have to be worked out by readers themselves. They have to bring together the visually available display, two curves of the same shape (width and height) shifted in location along the horizontal direction with 'directional selection', 'extreme phenotype', and 'favored over other phenotypes'. That is, readers have to find meaning through their informed reading of the lines alongside the text.

Here, selection is a process that holds like a natural law. The caption presents it in a context that allows the interpretation that it is a cause for the drift of the mean in the distribution. But there are three possible contentions. First, the graph is the result of some observation. The 'genetic drift' (natural selection) is an inference based on the graph. Second, it is the interaction of the organism with the environment that brings about the genetic drift. Third, 'directional selection' and 'stabilizing selection' are descriptors of the observation that variability decreases or mean property drifts; they are not causal determinants in the ecological site. That is, here as in many other cases, the textbook inverts causes and effects, and confuses observation with theory. This contrasts with the type of description that can be found in journal articles where a caption may have read something like 'a drift in the mean of the distribution with constant variability'. In the text, this would then be interpreted as an indication for directional selection.

The *main text* is not helpful in reducing ambiguity. First, we note that the first sentence of the text uses labels 'a' and 'b' for stabilizing selection and directional variation that are shown in figure 23.13b and 23.13c, respectively:

Figure 23.13 indicates that natural selection has two common effects: (a) to stabilize variations and (b) to direct variations. Stabilizing selection tends to eliminate atypical phenotypes and enhances adaptation of the population to current environmental circumstances, but directional selection selects an extreme phenotype better adapted to a new environmental circumstance. Directional selection occurs during a time when the environment is changing rapidly or when members of a population are adapting to a new environmental situation.[12]

Second, rather than providing a reading (instructions for reading), this sentence simply labels two types of effects. The following sentences then explain these effects, but no longer refer to the graphs. Third, the main text further enhances the confusion of cause (theoretical description) and observational description. Here, the instructions for reading the graphs suggest that stabilizing selection eliminates atypical phenotypes, and directional selection selects extreme phenotypes. The description becomes the cause; in the new conditions, 'extreme phenotypes' makes no more sense because they are the normal phenotypes, and as the graph suggests, are equally distributed around the mean as was the original population. Readers also need to know that the description 'atypical' refers to the tails in the displayed distributions.

There is also a confusion of model and data, and therefore between two types of scientific practices: induction and deduction. The first sentence of the main text suggests that the graphs provide an indication of two types of effects of natural selection, a typical frame for data based induction. However, there is no indication that the line graphs have been derived from data. Rather, they are models for the kind of distributions that would be observed provided the characteristic feature is measured on a continuous scale.

In many ways, this graph represents a polar opposite to that which we analyzed in the previous section (case 1). Whereas the scientific authors provided many resources in the figure, caption, and main text for reading the line graphs in a specific way, such resources were not provided in the high school textbook. Graphs which appear in high school textbooks violate the advice for design of good information sources: the less common ground that is shared between authors and readers, the harder communication becomes, and therefore the more needs to be said. The graphs in the high school textbook assume much knowledge about distributions, variability, and how to re-present them graphically. (In the Epilogue, we exemplify the extent of this knowledge by elaborating the present graph, caption, and main text so that they provide sufficient resources for the sense-making processes of high school students.) This assumption is unjustified. Furthermore, graph, caption, and text are such that they cannot easily be constructed as congruent. It remains with readers to construct, through multiple steps of reasoning, the relationship between textual statements and features of the line graphs and how they support or express each other.

Case 3: External temperature and metabolic rate in introductory biology

Our second analysis focuses on a graph with caption and associated main text (figure 2.5) from a high school textbook for introductory biology.[13] A first glance reveals that there are, in contrast to the graph in figure 2.4, a number of

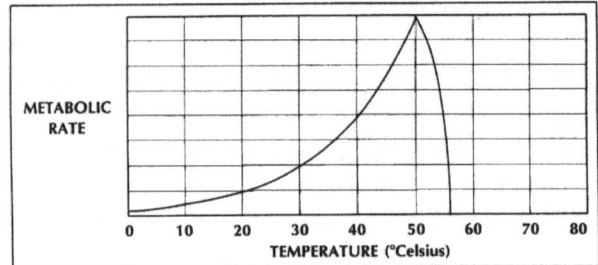

FIGURE 33-21. Metabolic rate is influenced by external temperature. Above about 50°C, the rate decreases rapidly because enzymes are destroyed. Most organisms cannot live at temperatures above 50°C.

Figure 2.5. Cartesian graph with caption from a high school textbook for an introductory-level biology course. (Quoted and reprinted [scanned graph] with permission of the copyright holder.)

resources provided: The horizontal axis features a label, units, and a scale. However, the vertical axis only indicates 'metabolic rate' as the variable, but does not provide units or scale. There is no source information as to the origin of the graph, so readers do not know if the graph refers to a specific animal, or represents animals in general. The smooth line, which 'calls for data points', and the scale on the horizontal axis suggest continuous measurement. But it is not clear what the unit of analysis is, nor if the graph re-presents actual measurements taken or a model that stands for metabolic processes in general. Interested readers who ask themselves what it is that is displayed, can find an explanation of metabolism about 570 pages earlier in the book: 'Chemical reactions within a cell both build up and tear down many complex molecules. The total of the reactions is called metabolism'.[14]

The first sentence in the caption provides a general description of the graph. It constitutes a causal relationship between external temperature (cause) and metabolic rate (dependent aspect). The verb 'influence' suggests a change in the dependent aspect with changes in the cause. In the second sentence, the caption provides both a description of the curve ('Above about 50 °C, the rate decreases rapidly') and an explanation for this shape of the curve ('because enzymes are destroyed'). The final sentence describes something that is not available in the graph. We cannot evaluate it with reference to the graph, because it does not show that 'Most organisms cannot live at temperatures above 50 °C'. The caption does not identify the unit of analysis or whether the graph presents collected data or a model.

The *main text* makes reference to the figure at the end of the claim suggesting that animals cannot survive in temperatures above fifty degrees Celsius.

Metabolic rate in animals varies with external temperature. Up to about 45 °C, metabolic rate increases two or three times for every ten degrees the temperature is raised. Approaching 50 °C, the rate slows down. Then it falls off sharply to zero. At high temperatures, the enzymes needed for metabolism are destroyed. Therefore, most forms of life cannot survive in temperatures above 50 °C (Figure 33-21).[15]

Through this placement, the sentence calls for empirical support. However, the figure cannot provide this support because, as discussed in the context of the caption, this 'fact' cannot be seen in the graph. That is, the figure is used in support of a generalization that it cannot support. Rather, this sentence constitutes an inference that could be based on this and similar graphs, and knowledge of metabolism more generally. The present text suggests that this is a fact that can be read directly from the graph.

As in our previous readings, the main text can be understood as providing instructions for how to read the graph. But when readers try to follow these instructions, they encounter trouble. The second sentence in the cited main text suggests that the metabolic rate increases two or three times for every ten degrees. But the competent reader of graphs will find that the factor of increase in the graph is exactly two up to fifty degrees Celsius. The text suggests that 'Approaching fifty degrees Celsius, the rate slows down', but there is no change in the rate of the displayed graph. A further confusion potentially arises because of the notion of 'rate' frequently used as rate of change in time-dependent relationships. In a study of college students' reading and interpreting of graphs, such interpretations clearly interfered with their construction of meaning from a graph that displayed birth rates and death rates as a function of population density.[16] In the present case, neither metabolic rate nor its derivative with respect to temperature (i.e., the slope of the metabolic rate graph) decrease.

If we look at this graph and the associated caption and main text from a critical perspective on scientific illustrations, we notice that it leaves open more questions and permits alternative interpretations than would be expected from scientific illustrations. This and the foregoing graph with caption and associated main text are rather typical for those that we observed in high school biology textbooks. Although this example provides a few additional interpretive resources for enacting a particular reading, these appear insufficient. In particular, there is a remarkable disjunction between graph, caption, and text; and the relationship between what caption and text say the graph displays and that which it does display is available only through a high level inference from theoretical frameworks which are not yet known to students. When actually present, the descriptions of graphs are often inaccurate or contradictory with what is depicted.

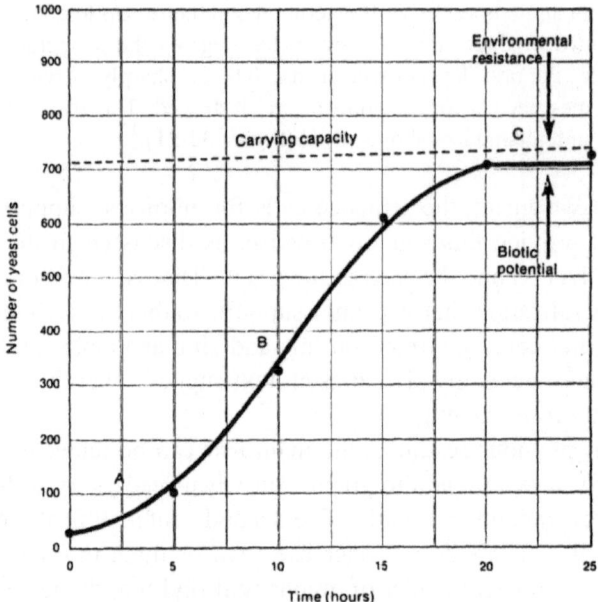

Fig. 38-2 Population growth curve for a yeast population

Figure 2.6. Cartesian graph with caption from a high school textbook for an introduc-
tory-level biology course. Graph and main text provide many resources for reading and
thus constructing meaning. (Quoted and reprinted [scanned graph] with permission of
the copyright holder.)

Thus, high school students are doubly handicapped in learning to read graphs.
First, they are not provided with many of the resources for reading and interpreting graphs that are available to scientific audiences. Second, what is provided in
terms of text is frequently inconsistent.

Case 4: Population growth curves in general biology

Our final graph (figure 2.6) derives from a textbook for general biology students
with lower than normal reading levels.[17] We included it here because, as we
show below, (a) it is unusual for those in high school textbooks and therefore
constitutes a 'negative case' and (b) it is in some ways exemplary for how
graphs could be presented in high school textbooks.

    This graph from a high school textbook, in contrast to all the others we ana-
lyzed, includes many resources that facilitate reading and interpretation. Axis

labels, units, and scales allow readers to relate this graph to the world of their everyday experience. The letters 'A', 'B', and 'C' mark certain aspects of the graph (without other information it is not clear what these letters designate). There are additional labels ('carrying capacity', environmental resistance', and 'biotic potential') and pointers available to mark aspects of the graph. Readers may also notice that the data points do not fall onto the smooth curve.

The caption constitutes no more than a title. (For more on the use of captions, see chapter 4). When held against the captions of the predominant number of graphs, one notices that this caption does not make a statement about the relationship in the graph. Rather, it labels the graph as a 'population growth curve' and therefore orders it into a class of curves. The main text underscores this interpretation of the graph as representing a class of curves: 'Most natural populations follow a growth curve like this one'. Because of the lack of information in the caption, readers have to seek additional interpretive resources in the main text.

The main text begins by elaborating a specific experiment that ultimately leads to the data presented in the graph:

Population Growth Curves

Many scientists have studied population growth. One of these did her experiments with yeast cells (Fig. 38-1). She started a yeast culture by putting a few yeast cells in a suitable medium (See section 12.7, page 158). Then she counted the yeast cells in a certain volume of medium every 5 h for a total of 25 h. Then she graphed the results. This gave the population growth curve shown in Figure 38-2. Note that this graph has three main regions:

• Region A — The population size increases slowly.
• Region B — The population size increases more rapidly.
• Region C — The population size levels off and stays constant.

Most natural populations follow a growth curve like this one. Now let's look at some explanations.

Factors Affecting Population Size

Biotic Potential and Environmental Resistance In Region A the population size increases slowly. This is because the number of "parent" yeast cells is very low.

In Region B the population size increases more rapidly. This is because there are more "parent" yeast cells to make new cells.

In Region C the population size levels off and stays constant. Why is this so? All species have a tendency to reproduce. This tendency is called the biotic potential. The biotic potential tends to increase the population size.

However, other factors work against that tendency. These factors are called the environmental resistance. As the graph shows, population size levels off when the environmental resistance balances the biotic potential. The ecosystem is now said to be at its carrying capacity. What factors do you think make up the environmental resistance?[18]

Elaborating a specific experiment is what scientists typically did when they interpreted for us their own graphs that they had displayed in scientific articles.[19] Parenthetically, it refers readers to a section in the textbook where they can find instructions for doing such an experiment. The text specifies the unit of analysis ('yeast cells in a certain volume of medium') and the frequency of measurement ('every five hours') and explicitly marks the graph in figure 38-2 as the result of plotting the data. The text subsequently directs readers' attention to three regions in the graph labeled 'A', 'B', and 'C'. With the identification of different parts, the text also provides instructions for reading the graph: 'population size increases slowly', 'population size increases more rapidly', and 'population size levels off and stays constant' are instructions for finding regions in the graph for which they are also descriptions.

In the graph, these regions are not clearly distinguished. Readers are required to return to the main text that distinguishes the regions according to population increase. Again, population increase is not directly available in the graph. Here, it is clear that the regions and their descriptions have to be recovered through an iterative process in which the meanings of 'increase in population size' and 'Region A/B/C' have to be elaborated. What the reader therefore has to find is the increase as a function of time, that is, the changing slope with time or, in more mathematical terms, the time derivative of the displayed function. Because the interpretive horizons of all readers cannot be predicted, the outcome of these iteratively achieved elaborations are inherently underdetermined.

The second part of the main text cited here (after the title 'Factors Affecting Population Size') can be read such as to provide an integration of the data displayed with a theoretical description. Each description of a section is associated with a cause. In regions A and B the number of parent cells is associated with the size of the population increase. For region C, visibly different from the others, a different kind of explanation is offered both in the text and in the graph. Here, two counter forces are cited in the text and entered as descriptors and pointers in the graph: environmental resistance and biotic potential. For a more complete explanation, students might want to know why biotic potential and environmental resistance appear to operate only late in the development of the population and not earlier. It is also not evident why the carrying capacity should be a function of time as indicated by the sloped line in the graph.

From our analysis it is clear that the text provides a great number of resources in the graph and main text for constructing meaning. Pointers and labels in the graph allow readers to integrate the theoretical notions with the data that clearly were linked to their experimental origins. The line graph hints at variability and error in that it does not clearly go through the points. Missing, relative to scientific publications, are more elaborate descriptions (instructions for reading) in the caption that could support the construction of meaning from graphs and text, and support the multiplication of meaning that can be ascribed to inscriptions.[20]

The work of reading graphs in scientific texts and high school textbooks

Graphs are a paradigmatic emblem of science and are used to exhibit continuous relationships; they embody mathematization of phenomena. At the same time, they contribute to the constitution of these phenomena. Without the graphs, the 'phenomena' are often not visible. For example, the authors of the paper in case 1 made visible drying rates and soil differences inside and outside corrals in several juxtaposed graphs that 'support' the claim in their text. At the same time, this phenomenon was only made available through the graph. Thus, the phenomenon and the scientific graphing practice are mutually constitutive.

Our investigations show that there is the same number of inscriptions in high school textbooks as there were in scientific journals dedicated to ecological issues. However, the nature of the inscriptions was poignantly different. The predominant inscriptions in high school textbooks were photographs, naturalistic drawings, and diagrams, all of which are based on a pictorial metaphor; pictorial metaphors are persuasive and intuitively attractive explanatory resources that support positivistic claims to realism of scientific knowledge. Our hunch is that their use as a didactic tool further reinforces traditional scientific claims to truth; a full analysis, however, is yet to be undertaken. With respect to Cartesian graphs, scientific journals not only featured a significantly larger number, but they were also of a different kind: High school textbooks most frequently used graphical models without scales that were almost non-existent in scientific journals. On the other hand, scientific journals featured significantly more scatter plots, scatter plots with connected points, scatter plots with best-fit curves, scatter plots with mathematical models, and mathematical models.

Our fine-grained analyses showed that there were also qualitative differences between the graphs; that is, the graphs exhibit considerable different topographies. Cartesian graphs in scientific journal articles provide many resources that assist readers to construct the relevance of data points and line graphs. Furthermore, scientific journal articles provided, sometimes extensively, additional resources in the captions that direct readers' attention and interpretive efforts. Scientific main texts featured specific readings of the graphs that were clearly separated from the interpretations that linked the featured data to theoretical frameworks. Our analyses showed that the captions and main text in high school textbooks feature considerably fewer resources and almost never additional resources and specific readings. For example, the unit of analysis to which a specific point corresponds is frequently made very explicit in the scientific literature either in the caption or the main text: 'Every symbol indicates the result of one experimental session', 'measured and modeled red and near-infrared reflectances for 31 black spruce stands', 'the density [dependent variable] of older trees in corral and noncorral stands can be concatenated to create hypothetical survivorship curves'. In contrast, the unit of analysis is almost never identified in high school textbooks. Yet it is the recovery of the unit of analysis, the original thing or individual on which pairs of measurements were taken, with which high school students are least familiar.

High school textbooks and scientific journal articles also differ in the way they treat Cartesian graphs with data and those that feature models; that is, there is a difference between data which are used to ascertain or to induce some relationship and graphs which express a relationship that was deduced from a theoretical model. In scientific publications, there is a recognizable relationship between 'real' data and the theory that is expressed in continuous line graphs; behavior is reduced to a single law or set of laws. The clean line of the smooth curve is given authority by the fact that the individual measurements do not fall on the line, and frequently are associated with bars that indicate errors of measurement. Points suggest accuracy and statistical variation, whereas the smooth curve approximating them holds out the hope for a simple relationship that can be expressed in a mathematical equation. Smooth curves displayed alone are looking for data points. Reviewers in the scientific community often take graphs without actual data as lazy attempts at demonstration. But such graphs are exactly what readers of high school textbooks continuously face. This interaction between empirical data and relationships of theoretical nature is no longer available to textbook readers. Here, graphs are detached from empirical situations to which they might relate. But textbook authors never make it clear that the featured line graphs are used to express currently accepted models.

We began this investigation with the overarching questions, 'What practices are required for reading graphs in textbooks versus those in scientific journals?'

and 'Do textbooks allow students the appropriation of, and development of competence in, the graphing practices as they are required for reading scientific texts?' First, in our attempt to construct answers to these questions, we showed here that there are both quantitative and qualitative differences between graphs in the scientific literature and high school textbooks. That is, we identified discontinuities between graph-related practices in high school textbooks and scientific journals. These discontinuities parallel those visible in the graph interpretation practices: Even B.Sc. students experience considerable difficulties reading graphs and usually enact readings that remain referentially stuck; the discontinuity appears to occur along the amount of conducting and reporting field research rather than the amount of high school and university courses. The discontinuities in representation use are therefore not likely to allow the kind of trajectories of peripheral participation that play a central role in the transformation of newcomers to old-timers in communities of practice. Results from other studies in our research program further support this interpretation. We further showed that preservice teachers with undergraduate or graduate degrees in science showed no more competence in using graphs to analyze data than pairs of eighth-grade students (see chapter 3). Second, we found a striking irony with respect to the resources and practices required for reading graphs. Scientists with much training and experience in reading and interpreting graphs are provided with resources, specific readings, and explicit links to theoretical frameworks. Students, who are identified in the literature as lacking graphing skills, are not provided with these resources, readings, and indications as to the theoretical or empirical origins of the graphs displayed. Scientists not only have more training and experience than high school students but, ironically, are provided with more resources for constructing specific meanings from graphs than high school students receive while reading their biology textbooks.

In the foregoing discussion, we highlight the differences between graphs in high school textbooks and scientific journals. However, this discussion has still left open questions about the nature of the work of reading graphs. In our phenomenologically informed anthropology of graphs, reading graphs is neither in the text nor in the reader. Reading graphs constitutes a social practice (phenomenon) that is known through its achievements and use. It emerges through readers' competent enactments of the community's ways of reading, in reading what the graph says. The graph and associated captions and main text provide the material grounds through which the practices of graph reading are organized.

Here we were concerned with biological texts. Biology provides ways of looking at nature. Biological texts, therefore, both in research papers and textbooks, serve as demonstrations in which readers are shown, and learn to see, the order reflexively specified by the textual arrangement of words and graphics. The text, then, is a worksite where knowledge is constructed not through the passage of information from the book or article to the reader, but through the recovery of the order encoded by the author in text and accompanying graphs. It is through the noticeable work of reading that the knowledge of biological objects is constructed. The convincing text engages readers in a process of authentication so that readers recover what the biologist previously saw. Texts that permit readers to achieve this authentication, that is, texts that scaffold readers in achieving the demonstrable fit between graphs and text, display a compelling pedagogy in teaching readers something about the world they inhabit.

Graphs do not simply display relationships, adjuncts to text, or proofs for something stated in the text that readers can 'take in' in an unproblematic way. Rather, to understand the biological phenomena of which the graph is a part, readers must engage in a reflexive elaboration in which the main text and caption provide iterable instructions for where to look and how to read the graphical display; they provide organizational resources to the reader's gaze and interpretation. The reading of graphs and perceiving of graph features is further driven by a person's current interpretive horizon, competencies in the textual practices of the scientific community, and familiarity with the domains to which graphs refer. This horizon provides motives for regarding, disregarding, and selecting among potentially important aspects. We further develop these points in chapter 4.

Signs such as words and graphs are present-at-hand in the totality of their relations to other signs and things. That is, signs do not have meaning on their own, but obtain it through their relation to other signs and in relation to readers' shared experience of the world. Sense comes from shared use. This sense is not available if we focus on the sign *as* sign, rather than using it transparently to refer to its referent. Therefore, when students do not reconstruct natural phenomena from the individual dots (graphs), we see this not as a sign of their 'inability' or 'cognitive deficiency', but as a lack of experience participating in the sociocultural and cultural historical practices of using graphs as a part of reading and producing such demonstrations. When graphs are used transparently as signs that refer to patterns, or resources to construct biological aspects, they are ready-to-hand transparent tools (like the hammer to a carpenter). When readers do not see the referent transparently through the graph, they have to attend to the graph itself. The graph is no longer ready-to-hand, but becomes present-at-hand like any other broken tool. It draws attention to itself, so that readers' concerns are now with the sign. We do not 'grasp' the sign when we stare at it, seeing it in its full sign nature. In cases with a high degree of intersubjectivity between interlocutors, a simple squiggle on paper, on white board, or in a hand gesture is sufficient to communicate the nature of a relationship. The resources normally supplied with graphs are no longer necessary to understand the nature of the referent. The resources do, however, assist readers not familiar with a particular project in constructing the object.

3   Graphicacy and context

In the past, researchers thought of (mathematical, scientific, etc.) literacy in terms of sets of skills that knowledgeable individuals took with them from one setting to another. Schooling, which takes place in special buildings separate from other activities in the community, is inherently based on this idea. You acquire the sets of skills during the first part of your life, and then apply it once you get out into the 'real world'. There is an assumption that schools are not the 'real world', but some other kind of world. Nevertheless, the literacy one acquires in school is supposed to be relevant to the different world where it is to be applied. These assumptions have been questioned by ethnomathematical studies designed to compare levels of schooling, competency in school mathematical skills, and mathematical knowledgeability in everyday activities, for example, enacted by grocery shoppers, child street vendors, street corner bookies, and dairy workers.[1] These studies showed that there were significant differences in performance on school mathematics tasks and mathematical practices in everyday settings, and that there was little or no correlation between level of schooling and the competence displayed in these practices. We can think about these differences in terms of context. In schools, graphical literacy is taught as if it was independent of the intents, emotions, structures, and politics of the situation in which it is used. In everyday out-of-school settings, however, the graphical literacy displayed is bound up with the purposes and interests of people, social relations (e.g., 'illiterate' Brazilian children draw on social relations to mediate their graphical practices), and material structures in the environment (e.g., dairy workers reason in terms of milk carton arrangements on shipping pallets). That is, context mediates levels of graphicacy.

Educators increasingly recognize the importance of context to knowledgeability. Typically, mathematics educators will address this issue by providing word problems as a way of introducing context. Science educators use everyday examples as part of their presentations to show how some scientific concept applies to the students' world outside school. For example, the physicist in chapter

1 attempted to help his students understand position-time, speed-time, and acceleration-time graphs by talking about earnings, hourly wages, salary, and salary increases.[2] While we recognize such approaches as earnest attempts in assisting student learning, the fallacy is that reasoning verbally is the same as reasoning in practical action.[3] Interestingly enough, the physicist did not so well in his attempt of translating the relationship between the physics concepts and the salary example. He likened a constant acceleration, which means going faster all of the time, to a constant annual salary rather than to a constant increase in hourly wages. These considerations raise the question, 'What is the role of context in knowledgeability?' Here we are concerned with the attendant question, 'What is the role of context in graphicacy?' and, even more important, 'What is the role of context in *critical* graphicacy?'

In this chapter, we analyze the role of context in students' interpretative practices concerning inscriptions, that is, their graphicacy. Therefore, graphicacy-intensive activities were observed in the same setting (school) but in two different situations during a ten-week ecology curriculum for eighth-grade students: (a) field research that required the production of inscriptions that convincingly supported any findings and (b) word problems that were in the form of stories about eighth-grade students who conduct field studies and that presented data in the form of student-produced inscriptions (well-known *con*text). This design varied the situation of graphicacy while controlling for students' familiarity with the situation described in the word problem; the use of student-produced inscriptions as part of the problem presentations and the maintenance of collaborative groupings in both situations were further attempts to make the problems and this study ecologically valid.

Context and interpretation

At the time of the study, eighth-grade students were engaged in a ten-week ecological study of the school's campus, a fifty-acre lot with a variety of ecological zones.[4] The goal of the ecology unit was for students to develop competency in designing independent research studies, making sense, and supporting findings in a convincing manner. This provided for many opportunities to observe graphicacy: applying statistics in the phenomenal world; translating between inscriptions; making connections between mathematics, the phenomenal world, and science; and solving problems and modeling natural phenomena. During the ecology unit, each student group was asked to find out as much as they could about their own small plot of land, called an ecozone, about 35–40 m^2 in size (figure 3.1). From these activities, and in order to construct convincing represen-

Figure 3.1. Jamie (left) and Mike are in the process of collecting data in the site that they had selected for conducting several studies with respect to plant and animal ecology. Here, they are in the process of measuring the amount of light that mayapple plants receive, which they will subsequently correlate with plant height and growth.

tations, there arose many opportunities for mathematizing their observations. The development of students' graphicacy was a core goal of the unit.

To maximize learning during the open-inquiry field study, the teachers imposed the following conditions: students were (a) to produce defensible arguments for the necessity of a research question and the soundness of the experimental design and (b) to present their results by means of inscriptions that would convince peers and teachers. In this, the learning environment incorporated one of the fundamental principles of scientific progress: Scientific practice has rhetorical aspect recognizable in scientists' efforts to support their positions. That is, students were held to exhibit and develop practical graphicacy. More so, because they had to defend their inscriptions and critique the inscriptions of others, the conditions existed for students to evaluate the relative merits of inscriptions. That is, there existed opportunities not only to develop graphicacy, but also to develop *critical* graphicacy. There are indications in students' conversations in the field and during the interviews that they were attuned to this critical attitude. They asked each other questions such as, 'How do you know that that's true?', 'How do you support your claims?', or 'Put lots of measurements, because [the teacher] likes people to prove stuff'. For example, back in the classroom, Jamie (one of the eighth-grade students) studied his notebook with the sequentially recorded data from his second field study. He suddenly exclaimed, 'Look at this one; the higher the pH level, the higher the plant, but the lower the amount [of increase]. The taller a plant, the lesser is the amount. The lower the pH, the lower the height, but the more the amount'. He proceeded to note this statement as a claim in his report. But about five minutes later, he urged his partner Mike that, to support this claim, they had 'to make a table which has to make a comparison which says, which compares the population density with the pH level and the growth'.

In the course of the year and to prepare the students for their ecology study, the teacher progressively shifted the responsibility for learning to students. During the first unit, on physics, he asked students to conduct specific experiments to answer equally specific research questions. During the subsequent chemistry unit, students designed their own experiments to teacher-determined research questions. The third unit, on ecology, was characterized by student-designed research questions and investigations. Whereas all students evaluated the freedom during the ecology unit very positively in the end, a few had problems getting started and designing their first research question. Some were overwhelmed by the extent of their observations; others found too little to see in their ecozone. There was much independence, and the performance criteria were not as clear as in their other, lecture-style classes: Students were, for example, asked to 'Support [their] claims in a convincing manner'. Thus, some students did not know what to do and expect or where to start. However, the supporting structure described next, allowed students to gain confidence and to adjust quickly to the new learning environment.

Typical fieldwork

Jamie and Mike wondered what pattern they could find in their data collected to answer the question, 'What is the relationship between soil acidity and plant growth?' They were trying to make sense of their data out in the field, so that, if they needed additional or different data, they could collect these.

01	Jamie:	We have to figure it out right now, the pattern.
02	Mike:	If you look at it though that's the [area with] low [pH]. It starts of, that's kind of, that's the [area with the] lowest [pH], that's the [area with the] highest pH, and that is the lowest, and if you look at this, it has this is the highest pH.
03	Jamie:	So the pH is chosen.
04	Mike:	The closest to neutral eighteen point nine [cm average height], then the one [area] with the lowest pH has twenty-one point five [cm average height]; it has sixteen May apples; then the one with the medium one has twenty cm. So it looks like the pattern.
05	Jamie:	Looks like the pH.

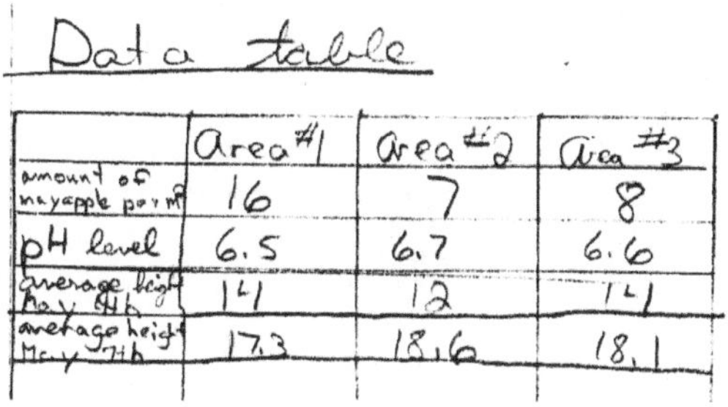

Data table

	Area #1	Area #2	Area #3
amount of mayapple per m²	16	7	8
pH level	6.5	6.7	6.6
average height May 6th	14	12	14
average height May 7th	17.3	18.6	18.1

Figure 3.2. Jamie and Mike assembled the data about the relationship between pH and mayapple height in a data table. The table includes the number of plants measured in each of three sampling plots and the average heights on two different days, which allowed them to make inferences about how pH might affect plant growth.

06 Mike: It looks like the pH, the lower the pH.
07 Jamie: The higher it grows.
08 Mike: Well, it doesn't really mean, like we could express it in that way though; because I bet if it had a one pH [pH = 1], it wouldn't grow at all. It would just be the acid. So you can't say, so there is no way that you can say, the lower the pH is [the more growth].

This conversation was typical of the ones in which students of all ability levels engaged during their fieldwork. It is characteristic in that students engaged immediately in making sense of their measurements and the patterns they could construct. On the basis of their three pairs of measurements, they came to the conclusion that there was an inverse relationship between pH and growth (speaking turns 06–07). However, Mike immediately saw a problem with such a statement, because plants would not be able to grow in soil with pH 1 (turn 08). After Mike and Jamie returned to the laboratory, they transformed their sequential records from the field notebooks by constructing a table of ordered pairs. They followed up on this investigation and in the course of two weeks constructed highly elaborate, multivariate relationships. These relationships were in terms of multiple independent and dependent variables and presented in tabular and graphical form. Mike explained his findings to another classmate during a sharing period:

The higher the pH level, the higher the plant, but the less amount it went up [smaller increase in height]. The higher the pH level, the higher the plant, but the lower the amount

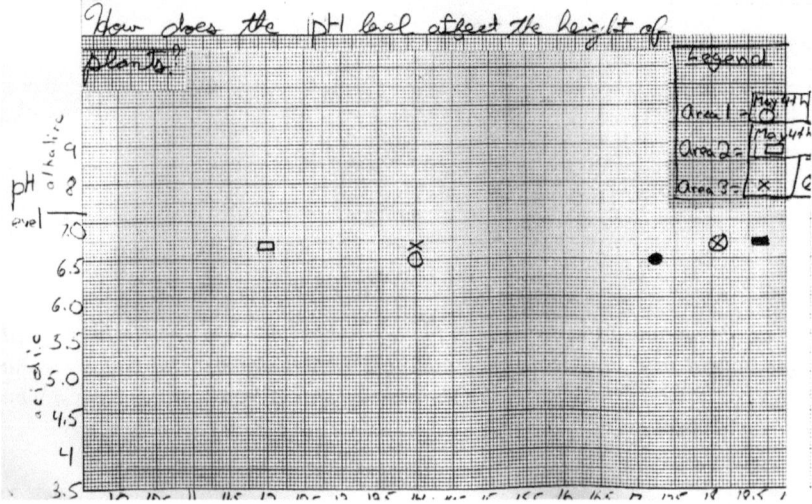

Figure 3.3. Jamie and Mike produced a graph in which pH and average plant height are plotted. They conclude that the largest plant sizes occur near neutral acidity (pH = 7.0), and smaller average plant sizes elsewhere.

[smaller number of plants]. The taller the plant, the less the amount. The lower the pH, the lower the height, but the more the amount.

While working in the field, Mike and Jamie had the opportunity to structure their situation. They defined and identified aspects of the environment that they thought might correlate. They decided which aspects were important to them, researched these, and reported their findings. In this way, over the ten-week period, they constructed complex understandings of ecozones and the relationships of biotic and abiotic variables. The present example was interesting in that the two supported their first claim with a table of ordered pairs. With the larger set for the second claim, they used graphs in addition to their table. That is, they had developed a level of graphicacy that may be more characteristic for high school or university students.

The videotapes illustrated a remarkable similarity between the (scientific) practices of the eighth grade students and those of members of a scientific expedition in Brazil. In both instances, members' practices transformed nature in such a way that it became accessible to mathematization. Such transformations are blackboxed when measurement instruments such as thermometers, pH meters, or anemometers are used. These transformations became more evident in other cases. For example, some student groups took soil samples back to the lab. By means of a sedimentation test (based on the differential floatability of materials), they separated each sample into several components and made scale drawings of the layers. Encouraged by their teacher, students then took rulers to determine the relative contribution of each component (in percent). At this point, the phenomenon was mathematized and could be subjected to further mathematical practices. Some students entered the data into tables; others plotted them so that they could compare their results with a published classification scheme found in one of the resources. Here, students' graphicacy was embedded in an entire series of practices that transformed nature into various forms of inscriptions. Some of these transformations had more material character (taking sample from ground, floating soil), whereas others had more formal character (plotting percentages). That is, these series of practices constituted a context for graphicacy, and, we might hypothesize, that the requirement to defend and critique inscriptions and the associate levels of emotional stakes was a central component of this context.

Where is the context for an interpretation?

During the research project, the 'Lost Field Notebook' problem was designed to answer two questions: 'To what extent do students use graphs in their fieldwork?' and 'Is there a relationship between the number of data pairs and the type of inscription used?' At that time, the teachers believed that mathematical problems embedded in stories about the students' own science experiences constituted ecologically valid testing situations; that is, the stories, because they described the eighth-grade students' experience, should engage the same mathematical practices. The stories were motivated by actual events in these classrooms (a lost notebook or a difference of opinions about the meaning of data). To increase further the likelihood of correspondence between our stories and students' experience, the story problems also included student-produced inscriptions to present data (the map with data in figure 3.4 was produced by another eighth-grade student). The working hypothesis stated that the students are more likely to use graphs (closer to form) with large numbers of data pairs than with small number pairs (closer to nature).

To allow comparison, the 'Lost Field Notebook' was designed with three data sets (one with the original student data, the other two synthetically generated) that were distinguished by the number of data pairs and the color of the sheet that students received. The three data sets are graphically represented in

The Lost Field Notebook

Erica is a grade 8 student doing research on an ecozone. She wants to find out whether there is a relationship between the density of brambles, a plant with a long narrow stem, and the amount of light these plants receive in different areas of her ecozone. She subdivided her ecozone into smaller areas. In each area she measured the approximate coverage of the area (in %) by the brambles. For each area she also found the average amount of light, measured in foot candles (fc). She recorded her data in her field notebook in the form of a the map reproduced below. Erica lost her field notebook, and you found it. You wanted to know the patterns she had found, but besides the map there was no additional information. Based on the information provided, (a) what patterns, if any, do you see? (b) what claims would you make? And (c) how would you support your claims?

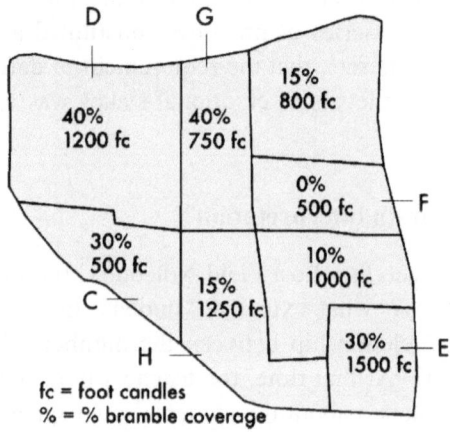

Figure 3.4. The text of this word problem is a vignette describing students' concurrent fieldwork. The map is identical to that submitted by one of the students as part of his fieldwork; the letters in parentheses have been added for ease of reference.

figure 3.5. The problem was chosen for three reasons. First, its apparent correspondence to students' experiences during field research was strong. Second, the problem in its three forms is equivocal even for individuals much more experienced in research (e.g., graduate students); this equivocal nature promised rich debate and therefore occasions for exhibiting *critical* graphicacy. As indicated in figure 3.5, the consideration of various points as outliers improves the statistical fit of linear regressions. For the data set plotted as squares, considering point 'A' as an outlier, or using a quadratic polynomial yields a good fit. The

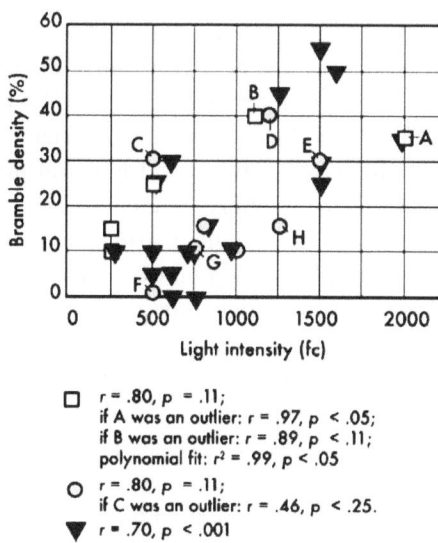

Figure 3.5. Plots of the three data sets and regression statistics. Because of the scattered nature of the data, univocal interpretations are not easily achieved, leading to a great potential for lively discussions among students.

second data set (circles) with eight data points changes from a non-significant to a significant relationship when point 'C' is considered an outlier. The data set plotted using black triangles was interesting because novices might consider the scatter too large to support a relationship. Third, from a didactic perspective, the problem was interesting because it shared similarities (in the scatter) with data sets that practicing natural and social scientists encounter in their work. As such, this problem deviates from most traditional practices in schools where students are provided with data or produce data that show unequivocal functional relationships.

Although the students easily framed problems and enacted knowledgeable graphicacy during their fieldwork, they reacted differently when faced with word problems such as the 'Lost Field Notebook'. Here, many felt that 'You cannot solve the problem if you do not know where the ecozone is'. Whereas they would normally proceed to the field, collect data, and transform their measurements in various ways to construct convincing presentations, they initially appeared at a loss with the present task. There appeared to be a consensus that the teachers had not provided enough information. For example, students suggested data and interpretations would be significantly different if the data had been collected near a pond, lake, or nuclear power plant. Some students asked us for the size of the area displayed, how the area was oriented (N–S, E–W), and what the soil moisture readings were. Localized low soil moisture could affect plant growth producing aberrant readings. Students also asked the teachers if there were trees either surrounding or in the ecozone, and what the extent of the canopy was. As Kathie indicated, 'trees might be affecting the amount of foot candles, and they are blocking out the light, and that affects the growth of brambles'. Other students thought that, if they had found the sheet, they would go back to the field site and study the entire area and each tile in particular to find out how many trees there were and how they shaded each tile.

To make sense, students wanted to 'invent another factor', 'assume that straight up is north', or 'assume that the data had been taken on the same day'. There were also a few suns drawn onto the map of the problem statement, which can be interpreted as indicating directions and most probable lighting (see figure 3.7 below). A few students suggested that there might be typographical errors in the problem statement itself. One student suggested that if he had found the notebook, he 'wouldn't even care what the pattern is'.

Although the 'Lost Field Notebook' was designed as an ecologically valid measure of students' graphicacy during their fieldwork, their reactions made this assumption problematic. Rather than experiencing the task as something that was similar to their field research, students found that it had limited their access to information that they usually obtained in the field. This information was necessary to construct reasonable patterns in their data, which they then presented in various inscriptions. In spite of students' questions, we insisted that they attempt to provide solutions.

Of the nineteen groups that were tested, eleven had used some form of mathematical inscription as part of their verbal presentations, whereas eight groups made verbal presentations only. Among the mathematical inscriptions were seven graphed solutions, two tables with ordered values, two answers that resembled tetrachoric correlations, and three pattern maps (one group produced three inscriptions). Ten groups explicitly discussed variables other than light intensity that could have affected plant density and thus prevented a clear pattern from emerging from their data set. Seven groups suggested measurement or transcription errors as possible sources for the 'irregularities' in the possibly existing pattern.

These results were especially interesting when contrasted with the results recently obtained in a methods course for preservice secondary science teachers in a five-year program. Here, only one of seventeen students, all of whom had previously earned B.Sc. degrees in science or mathematics, proposed graphing, whereas the other sixteen did not use any mathematical transformation and suggested that there were no obvious patterns. These preservice teachers generally agreed that the 'Lost Field Notebook' was too difficult even for high school students, let alone our middle school students. We have asked a second group of preservice teachers to do this problem. Of this group, fourteen had obtained a B.Sc., four a masters of science degree, and one a Ph.D. in a science (biology, chemistry, or physics). Only four preservice teachers provided a solution that included a mathematical inscription, whereas the others argued (without supporting inscription) that there were no patterns.

The analyses revealed that having previously made use of a specific inscription was not a sufficient condition for making use of the same inscription on this task. Seventeen groups of eighth-graders had at least one member who had previously used graphs. However, only seven groups actually produced a graph. Similarly, tables had been used by at least one member in each of the nineteen groups, but only two groups used a table with ordered pairs as a target inscription. Averaging procedures to compare subsets of data that had never been used to this point in the course were spontaneously 'invented' by two groups; similarly, the use of maps was invented in one group but constituted a repeated practice in two other groups.

Mathematics achievement (as measured by their grade in mathematics) predicted the students' choice of inscriptions. Those groups with higher mathematics achievement were more likely to transform the data into a graph, ordered data table, or pattern map; groups with lower mathematics achievement were more likely to provide an answer based on inspection of the map alone. When contrasted with our earlier findings of no predictive quality of mathematics achievement on choice of inscription during field studies, this result adds some support to the claim that, despite our contextualization efforts, the present task was more like a traditional mathematics textbook problem. The process and product aspects of students' solutions are discussed in the next section. The problem presented to the students provided more than just an occasion for exhibiting knowledgeability with respect to the inscription at hand, the map and number pairs; the problem provided opportunities for producing inscriptions and therefore for exhibiting *productive* in addition to *interpretive* graphicacy. We begin by discussing those solutions that used one or more inscriptions to transform and order the information given by the story problem; in the next section, we deal with those solutions where students did not use inscriptions.

Productive graphicacy

Figure 3.6 shows one of the seven graphed solutions. The two students proposed two alternate best-fit 'patterns'. The accompanying text read:

There is a pattern from the data given. Where there is greater light, the percentage of plants is greater except for one point, which is totally opposite of that. We have picked a pattern. But it could also go toward the outlier, which would make the other point an out-lier.

In all the data (which is proved in the graph), where there is more light, there is a higher percentage of plants. This could be because, with all the sunlight, the water evapo-rated, which kills the plants. If sunlight increases and the water increases, the number of plants in percent would increase. In the graph, the optimum amount of light is noted.

Theirs shares with two other solutions that the dependent variable, plant density, was drawn on the x-axis (which is in disagreement with the conventions of the scientific community). The students explicitly marked one point that they con-sidered as a possible outlier. They recognized that the pattern would change if the other point were the outlier. This solution also exhibits another feature that, from our experience, is rarely found in mathematics courses: The students made sense of the graphs by relating the pattern to possible features in the environ-ment. In this example, students interpreted the right-most point as a maximum density; they further suggested that an increase in the light intensity would evaporate too much water and thus kill the plants.

Some of the rationales that students provided for using a graph as a con-vincing inscription included, (a) 'The only way we find a pattern is on a graph'; (b) 'If you want to find out something, then you graph it, and make it easier. Be-cause when you have a trend line, it is easier to find patterns'; (c) 'As soon as you said "how are you gonna prove this?", I was thinking of a graph'; (d) 'I started to convince him that it's giving us a better idea'; (e) 'It looks a lot better, and it helps you to find the pattern'; and (f) 'A picture says a thousand words'. These statements reveal two aspects of graphs that had become important to the students. First, graphing is a practice (for some, the only practice) that allows one to discern relationships or renders the work of establishing relationships eas-ier. In this, of course, the students reflected the recognition that the ease of com-prehending graphs is much higher than that of raw data on maps or in tables: The graph can be read at a glance, whereas the other representations ask the

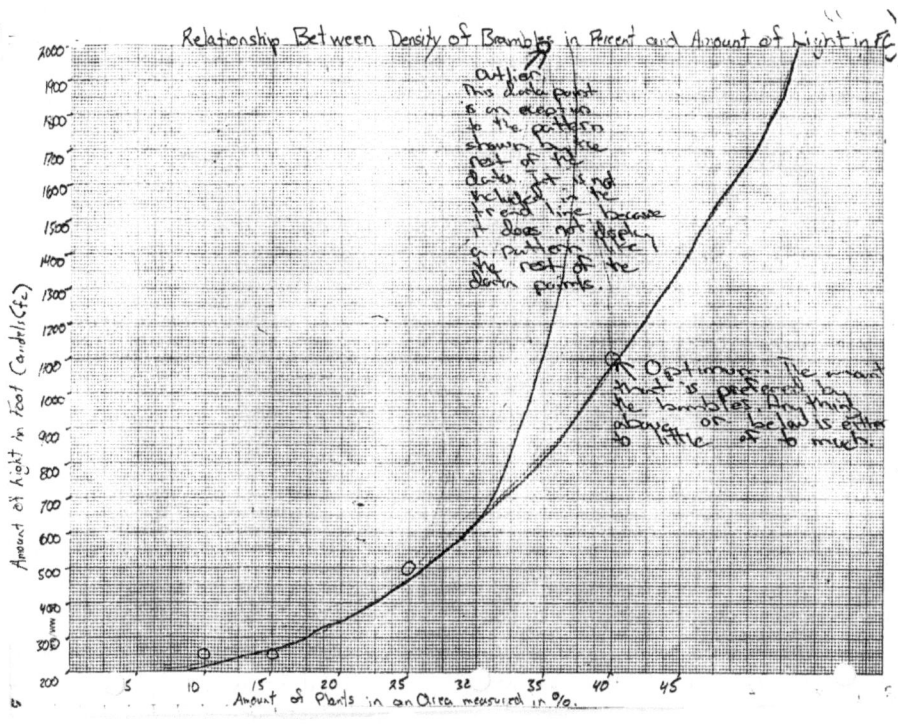

Figure 3.6. The interpretation by one group of students produced to the Lost Field Notebook. Two possible best fit ('trend lines') line were discussed, including the physical meanings of each.

reader to do the required arithmetic. The second aspect pertained to the rhetoric power of this inscription, for, as Jamie acknowledged, 'a picture says a thousand words'. In this perception, we see a result of the teacher's emphasis on using 'convincing' evidence and representations, which was based on our reading of recent work in the sociology of scientific knowledge.

By exhibiting productive graphicacy in their attempts to find patterns, students stripped the situation of even more physical context. Their concerns were, for lengthy periods of time, predominantly of routine mathematical nature. At issue were the nature of transformations to be made and the nature that trend lines would have. With the discussion of trend lines, the students began talking again about possible natural settings that their trend lines might describe. In this, they attempted to move from form to matter. The two students in the following excerpt entertained the possibility of two best-fit line graphs to describe the relationship between light intensity and brambles. The following episode occurred over and about the graph (figure 3.6).

01   Fabricio: See there is a lot of light; you might have a lot of light.
02   James: And water.
03   Fabricio: Because you might have the same amount of water everywhere and as soon as the light goes up, it might evaporate a lot of the water.
04   James: And dry out the soil.
05   Fabricio: And the plants would be getting more light, but they would also be getting, so I guess this would be the what is called the optimum light. And see there would then be, more light but it might be the, might not be; it might hurt them; that's the way it gets; it's too much; it has an amount that it can't go up to. So that might be it.

In this excerpt, Fabricio and James attempted to understand the relevance of a best-fit curve with a maximum. The teacher asked them why a graph might level off. They had no problem with establishing the positive relationship between light intensity and plant density. The transcript picks up where they tried to describe in their own words why this positive relationship might not hold for higher intensities and, in fact, decreased for one of their two best-fit options. As Fabricio addressed high intensities, James added water as a possible intervening variable (turn 02). Fabricio picked up on this idea and suggested a relationship between light intensity and the evaporation of moisture in the soil (turn 03); this, as James indicated, would dry out the soil (turn 04). Fabricio concluded that this might hurt plants, and therefore there was a certain optimum amount of light; beyond, there were no further benefits to plant density (turn 05).

In this excerpt, representative of those that we recorded in other groups, Fabricio and James collaboratively constructed an environment in which the data for the graph could have been collected. At first, Fabricio articulated the increase in light to which James added water. As he combined the two factors, Fabricio hypothesized that an increase in sunlight would evaporate much of the soil moisture. James rephrased this statement in terms of the soil condition: dry soil. Fabricio then concluded that, while plants would get more light, 'they would also be getting' less water. Although he did not state 'water' explicitly, his suggestion that more than the 'optimum light' might actually hurt the plants can be read in this way, for the two did not consider any other factors. Here, their prior experiences of changing soil moisture levels with changing temperature helped them to give the inscription more material character by transforming it into a possible ecozone. The shape of the graph received meaning as Fabricio and James engaged in a process of descent, from an inscription with formal character (graph) to actual, lived experience that reversed the earlier process of ascent, which had led them from the field experience to a graph relating soil temperature and soil moisture.

Two groups of students suggested a comparison of means. One of these made the following written argument:

We claim that the brambles grow better in large amounts of light. We can also prove our claims by averaging the amounts of foot candles of light being received by the brambles. The left side ((points to the number 1139 on his sheet)) averages more than the right side ((points to the number 790)). Now seeing that there is more light on the left side ((points to the number 28)) of the map, we can also see that the brambles grow denser than on the right ((points to the number 13)). Therefore it is evident that brambles grow denser in areas, which receive more light. But this is not the case always. As shown in the lower right corner, there is a very large amount of light entering the area, but not a high-density figure of brambles.

Here, students grouped tiles with low and high plant densities (median split). For each of these two groups, they calculated the average light intensities and plant densities. Subsequently, they compared the means of these artificial dichotomies and concluded that there was a positive relationship. They finally softened their claim by stating that individual cases may differ from their generalization. The other group argued similarly but had made a split of all the data into low-intensity (less than 1,000 foot candles) and high-intensity (above 1,000 foot candles) tiles. They noted that those areas with high light intensity, on average, had higher plant densities. One group calculated overall averages of light intensity and plant densities without making further comparisons. The process of collapsing interval data into two or more categories is not unfamiliar to many social scientists. The two groups here developed a procedure that was much like a tetrachoric correlation that correlates artificial dichotomies both on the independent and dependent variables. As classroom teachers, we were surprised by the sophistication of such spontaneously produced answers.

Two groups ordered their data pairs in tables. The group with five data pairs came to the same conclusion as the graphing group cited above: Plants like a mid-range of light intensity, between 900 and 1,300 foot candles. The second group had eight data points and suggested that large plant densities appear to be associated with high light intensities (above 1,000 foot candles). They concluded, 'light levels of 500–800 fc still have high percentages of brambles because 500–800 fc is a reasonable amount of light'. In the two cases, the students explained their on-going activity: 'I am just putting them in order, from least to greatest,

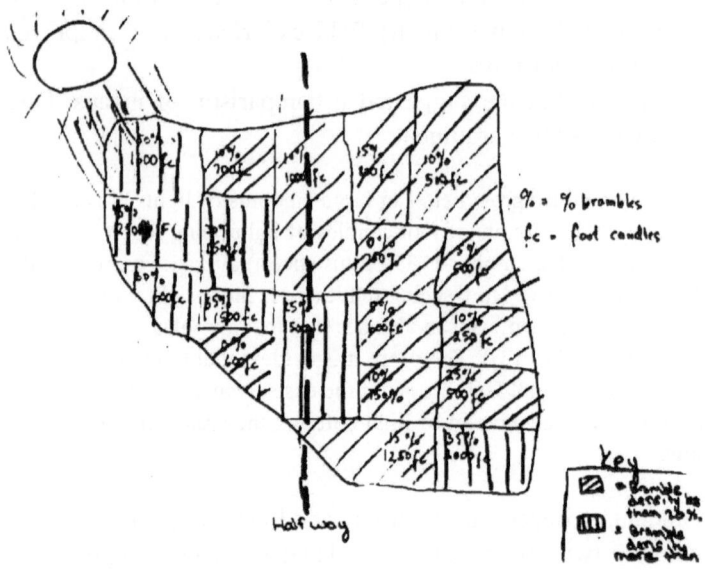

Figure 3.7. An example of the pattern maps used. The students used these to discuss trends in their data that relate to geographical distributions.

and I am trying to figure out patterns' and 'We try to chart our information to see if there is any pattern'.

Three groups shaded their maps according to the amount of light or plant density (e.g., figure 3.7). All plots with a density or intensity above a threshold value received one type of shading; all those below were marked in a different way. Incidentally, two groups then resorted to using the previously described averaging procedure. However, by shading, they were able to extract even more information than from the averaging alone. After concluding that there appeared to be relationships between light intensity and plant density, they hypothesized (on the basis of the pattern of the shading) other possible factors affecting the area, such as direction of the sun and shade trees.

Where there is more light, there is a greater percentage of brambles—some exceptions. There is a lack of sunlight—maybe caused by shade—on the right half of the ecozone. Since there is shortage of light, there is a smaller percentage of brambles. There must be a tree or something blocking the light.

Interpretative graphicacy

Groups who answered the questions on the basis of verbal presentations without resorting to some form of inscription often found it difficult to come to a consensus about whether or not their data could be expressed as a pattern. Four groups described patterns they observed in terms of the spatial arrangement of tiles with high and low light intensity or plant density without actually shading their maps, or other visual aids. Thus, students suggested among other things, that 'The sun is shining on a diagonal from the South-East; therefore the average foot candles will be the most where the sun shines on', 'The corners of the eco-zones have lots of percent, covered with brambles', or 'The western side has much more brambles but has not as many foot candles'. Although the intent of these verbal presentations appeared similar to that of the students who used the pattern map, the procedure appeared less generalizable; the map presented a visual pattern that could be apprehended in its totality, but the verbal presentation did not support a similar holistic process as the visual image.

Groups who used a graph, averaging procedure, table, or pattern map all suggested positive relationships between intensity and density. They focused their written presentations on elaborating the meanings of their inscriptions. The other groups developed their arguments verbally. Of these, four groups concluded that there were no relationships, whereas the others concluded that there were. The following two excerpts exemplify the two types of answers.

The higher the amount of light, the density is thicker. The more amounts of light cover more of the area and the less amount of light covers a smaller area. The plants prefer light and are more dense where there is more light. The areas that receive less light has more of other plants and not brambles, 15 % brambles and 85 % other plants in Zone 1. As light increases the density of brambles increases. When there is too much the number of brambles decreases again in Zone 3.

After looking at the map, it was noticed that there was no apparent pattern. There would be a pattern, except for the fact that there are too many exceptions to make a pattern. For example, Area 1 (see map) has 750 fc and has only 10% of brambles; and Area 2 has 500 fc and has 30 % of brambles. . . .

The students who produced the first statement did not address the scattering of the data points; rather, they indicated that the positive correlation changes for high values. The two students even re-drew the map and then entered the information in the way they saw it, including the percentage of area for other vegetation (figure 3.8). Most of the groups that suggested positive relationships between the two variables also suggested that the data were scattered because of

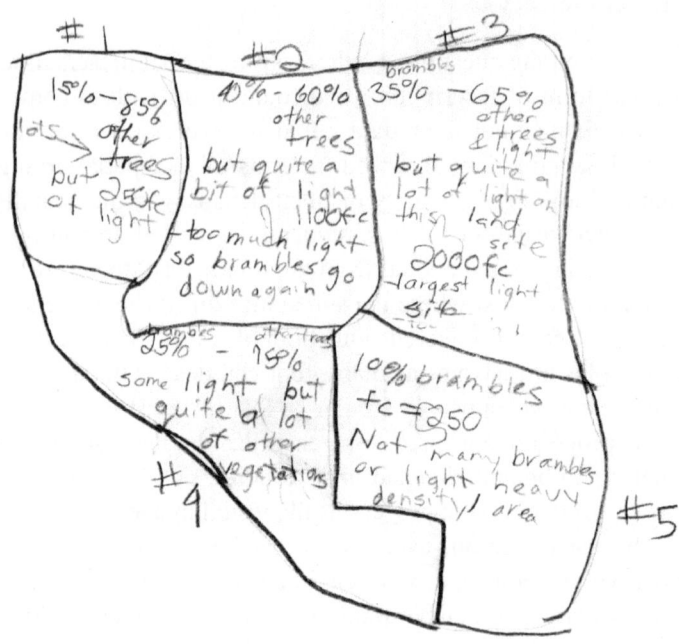

Figure 3.8. An example of the maps into which relevant verbal information was inscribed. The students used such maps as a way of transforming the data into verbal-geographic terms.

some other variable or measurement error. The second group saw enough evidence for rejecting a relationship between the two variables.

The following episode provides a glimpse at the reasoning that led Shaun and Lisa to their results. In his first field project, Shaun had done a similar experiment but had correlated only two levels of light (above/below 200 fc) with two median heights (1 1/2–2 and 3–4 inches). In the present context, Shaun and Lisa found themselves in the position that most of their peers had been in at the beginning of the course. By not using a mathematical inscription to summarize the information available, the task of coming to a conclusion was problematic. (Lisa and Shaun had this discussion over the map with the data. For ease of reading, the data points—'C', 'D', 'E', and 'F'—are referenced to figure 3.1 and 3.2.)

01	Shaun:	So the pattern would be, the more amount of light, the more amount of brambles, with the exception of one ((Points to C)) where the
02	Lisa:	No, no, the exception is this ((Points to F))
03	Shaun:	The exception is this ((Points to C)), because the more amount, the more amount of brambles, the more amount of foot candles
04	Lisa:	Yeah, okay. But what about this ((Points to D))? There is thirty percent and there is fifteen hundred ((Points to E)) and–
05	Shaun:	But it's still, it's still a lot, there is still a lot

As we had expected, the fact that the data were not neatly aligned to make a clear-cut decision provided for much discussion. In this case, Shaun held that there was a relationship between the amount of light and the density of brambles, and Point C was the outlier. Lisa (who generally had higher grades in all subjects) thought that there was no relationship and that Point F was the outlier (turn 02). She supported her argument by pointing to Plot E (figure 3.1), which was below 'D' (turn 04). Shaun hesitated but was not totally convinced by this argument. In the end, however, he submitted to Lisa's argument and the two constructed the 'no-pattern' answer reproduced earlier. Readers have to keep in mind that their discussion was not over and about a graph but the map. They used individual data points as evidence that the pattern proposed by the other was not universally valid in the data set.

The eighth-grade students did not rely on mathematical formalisms but reasoned directly within the situational context glossed by the story problem. Here, the students proposed a relationship and then used specific instances in which the hypothesized pattern was violated. Although the problems were of different content and structure, the same generate-and-test approach underlies students' reasoning in both situations. The groups who held that there was a relationship generally reasoned that there were a few exceptions to the rule.

The previous interaction is especially interesting because Shaun and Lisa were sitting near two other groups that graphed their data to find out if there were any discernible patterns. In the first instance, Shaun leaned over to observe Fabricio, who had just completed his plot.

| 01 | Fabricio: | ((To himself)) There is a pattern, except for this one right up here, except for that. ((Addressing Shaun)) See, it goes like this. And this, this is called an outliner, an outlier, right? |
| 02 | Shaun: | We guys have a different one, though. |

Fabricio muttered something about a pattern. He then addressed Shaun, traced with his finger the shape of a best-fit line that would describe their data, and designated 'A' (figure 3.5) an outlier. Shaun's rejection pertained both to the use of a graph and the particular best-fit Fabricio had produced because he had a different data set (turn 02). In a second conversation, Fabricio asked Shaun to support his finding ('How do you support your claims?' 'How do you know that that's true?'). Fabricio's and Shaun's groups had used graphs because, as later interviews revealed, they felt that they could (a) make the most convincing argument with this inscription (Fabricio had also considered a table with ordered values) and (b) identify outliers more easily. In spite of these two interactions, Shaun rejected the use of a graph. Following his partner, Lisa, he gave up on the idea of a relationship. In their argument, they used two sets of data pairs (C vs. G, D vs. H) that contradicted the correlation originally suggested by Shaun.

In this situation, Shaun exhibited *critical* graphicacy. Although his task looked almost identical to that on which Fabricio was working, he did not consider copying the solution. There were different numbers of plots on his map, which necessitated a different approach. Fabricio, too, displayed *critical* graphicacy when he asked, 'How do you know that that's true?', meaning the claim that his peer had made based on his representation. On the other hand, we illustrated that a number of Shaun's peers had argued for the rhetorical strength of graphs. They were thus already committed to those beliefs to which scientists (among them the teacher of the class) generally subscribe; accordingly, the inscriptions that are more universal in the cascade of inscriptions (maps for recording all the measurements, lists, tables, totals, means, graphs, equations) are more convincing than those that have more local character. That is, these students exhibited a considerable level of graphicacy; but if such graphicacy takes the form of ideology, it may assist students to cope with some situations but not with others where a critical evaluation of multiple forms of representing phenomena is called for. In this situation (word problem), Shaun and others did not support an argument with an inscription, although they had done so previously on their fieldwork. They made use, however, and some may say inappropriately here, of another successful scientific practice—finding cases for which a generalization is not true.

The reasoning of those students who rejected the existence of a relationship was exemplified in the following excerpt from our interview with Shaun. Both pieces of evidence (interview, artifact) are consistent with the transcript of his interactions with his partner and other students in the class. The transcript furthermore reveals that he had 'tried to do a little diagram', but this did not become part of the artifacts we received from him.

I was looking for basically the more light, the more percent of brambles or the less amount of light. Doesn't the pattern have to concern all of the point? That threw me off. I thought a pattern had to relate to all of them. Every single one had to be related somehow. Isn't that what a pattern is? Because of this one I sort of thought there is no pattern. I tried to do a little diagram of it, and because of this I thought a pattern had to be connected to all of them, and this one was off, so it was not a pattern.

This excerpt illustrates many students' beliefs about the nature of the universe. Much like most scientists, they assumed that relationships between variables are ideal. The graphs scientists use in order to communicate their understanding to novices generally are smooth, continuous, and give no indication of the 'real data' from which these idealized graphs were generated. The idealized graphs, in fact, are indicative of the mathematical structure presumed to underlie natural (as distinguished from social) phenomena. The students who 'disproved' the existence of patterns did so on the basis of two or three counter examples. Those who argued for a pattern in the data suggested and discussed other possible factors that mediated the ideal relationship they presumed. These factors included other variables, random errors, and measurement errors.

Students who had produced a mathematical inscription often talked about outliers without concerning themselves with the underlying causes. On the other hand, students who produced verbal presentations often entertained the possibility of other variables, random errors, or measurement errors inherent in their data set.

We can assume that another abiotic factor probably affects the percentage of brambles. Because of averaging the percentage of brambles and the light in each area of the eco-zone, the information gathered may not be accurate.

In the absence of other information, the group generated possible explanations about why an ideal relationship was not exhibited in the data. They suggested the possibility of other factors and questioned the measurements. The particular excerpt problematized the procedure of determining percentage brambles and light in the area; other groups suggested such systematic errors as temporal shading by surrounding trees at the time of measurement, or variations in the light measurement due to the temporal extension of the measurement process. As shown, the students' learning environment provided many opportunities to make the measurements themselves problematic so that the exact location or frequency of a measurement, the representative nature of a sample of plants for an entire area, or the number of measurements needed to represent 'average light intensity on a 1-m^2 tile' was a situated, negotiated, and not at all self-evident achievement.

Contextual mediation of graphicacy

The students in this chapter learned in an environment where they normally framed questions about the phenomenal world on their own; they collected their data and organized and transformed them using a variety of mathematical inscriptions including graphs, tables, pattern maps, comparison of means, summary statistics, or multidimensional analysis. After completing data analyses, they prepared reports with the intent to make convincing arguments for the importance of the questions, soundness of the data and transformations into various inscriptions, and relevance of their conclusions. Depending on their current knowledge and the requirement to argue convincingly, the levels of graphicacy students exhibited implicated inscriptions of varying formal character.

The present chapter concerned inscriptions presented as part of word problems, a minor aspect of students' work during the ecology unit. Here, the students' task was to make sense of an inscription rather than the phenomenal world. This inscription was part of a traditional story problem. However, the stem was in the form of a vignette that described the students' concurrent field experience; one of the three data sets accompanying the text was an identical copy of a student-produced inscription; the other two constituted variations.

At first, students searched for more information that would make the current setting more like their field work, which allowed them to situate their meaning making in a larger context. From the students' perspective, the text limited their options for constructing meaning. The assumption underlying all students' initial work was that relationships had to be unequivocal, that any deviation from a (curvi-) linear relationship was caused by some other, intervening variable. All groups realized almost immediately that there were no such relationships in their task. Here, they began to ask for more information; this quest for more information was so widespread that it began to exasperate the teachers present. Although the problem had been for the teachers a simple 'number game', a matter of using one inscription or another for investigating and displaying relationships among the data, the children were asking for information about the context in which the data had been collected. By abstracting the problem from the environment, the teachers had limited the range of options students had available during their fieldwork. The teachers had changed a meaningful setting into a puzzle with few options. The natural environment in which they did their field research allowed them to check for other variables, to develop measurement procedures as contingent achievements that arose from the social and physical interactions in specific settings. In the context of their field research, students appeared to have available resources that helped them deal with the questions they had framed, which therefore were central aspects of the graphicacy that they displayed. There, the data they collected were part of a rather complex situation that the students learned to structure. The inscriptions in the 'Lost Field Notebook' were, as the students said, 'just words and numbers'. Their meaningful experience doing research had been turned into a traditional school exercise. This finding is consistent with other research that showed that students' graphicacy in the context of purportedly realistic word problems often differ significantly from those on out-of-school analogies; it is too simplistic to think of word problems as lying on a continuum between abstract and concrete. Word problems as text share with instructions that, in order to understand them correctly, readers have to share the author's common sense; this would allow them to understand all that which goes without saying.

After the initial shock, and upon the teachers' (naive) insistence that they try and make do just with the information given in the inscription, the students used their past experiences in the field as a resource to deal with the apparently conflicting pieces of information on the problem sheet. The fact that their graphicacy (mathematical inscription or verbal presentation) in the context of the present task (but not their field work) could be predicted on the basis of their concurrent mathematics achievement, whereas their graphicacy in field studies could not be predicted, is additional supportive evidence for the claim that students approached this task in a way more characteristic for traditional word problems. This suggests the problematic nature of the assumptions that underlie many traditional studies on transfer. In the present situation, students, even academic high achievers, did not simply transfer their graphicacy from the field to word problem situations. Rather, once students had begun to discuss their presentations and inscriptions, they used their field experience in various ways to make sense of the data; that is, they rebuilt context such as to imagine environmental settings that would give rise to their specific data set. Of course, there is a large if not unlimited number of such environments, for any inscription that stands for some phenomenon essentially underdetermines the phenomenon.

The students' interpretive work related to map and numbers (figure 3.4) presented here underscores the sophisticated nature of their graphicacy while doing story problems. This assessment received further support when we contrasted the eighth-grade students' work with the responses from college students with undergraduate degrees in science or mathematics. The eighth-grade students' responses can be considered as the beginning of more interactions at the classroom level that would allow graphicacy to become a collective issue, and, thereby, turn into *critical graphicacy*. The teachers of these eighth-grade students did not take the classroom discussions any further because their goals as science educators were different and the mathematics department showed no interest in integrating the two subjects. Although the teachers had failed to take the lesson further, to turn it into an experience of collective inscriptional activity and graphicacy, they recognized its potential for informing future curriculum development that allows for the concurrent study of mathematics and science.

The students' interpretative practices in the 'Lost Field Notebook' situation allowed them to draw on their experience from the field studies situation within the course, so that the levels of graphicacy displayed involved some (to be determined) level of contextualization. In most science courses, however, they do not have such opportunities. In fact, word problems are usually written to conceal rather than reveal; they are more like puzzles and obstacle courses. Despite their experience in actual field settings, of which the 'Lost Field Notebook' was a vignette, and despite the nature of the data provided (a student-produced inscription), the task students faced appeared to be more similar to their normal mathematics experience than to that during their field work. One has thus to question the use of traditional contextual word problems. Students often have little to construct *con*-text, that is, that which normally goes *with* the text to make it meaningful. Besides concealment, there is another aspect that makes it difficult for students to make sense of a word problem's text; the common-sense knowledge it takes to understand (any) text is that which the writer leaves implicit because it goes without saying. In their field setting, our students framed their own problems, bringing to bear their own common-sense knowledge, and answered their questions accordingly. The inscription in the 'Lost Field Notebook' implied *the teachers'* common sense, likely different from the students'. The latter then resorted to another type of common sense: viewing the task as one that they traditionally faced in school.

These results cannot be understood apart from the special setting in which it was conducted. In the past, researchers focused on the type of inscription and the correctness with respect to the standard solutions produced by individuals. The present study was conducted in a classroom where students were used to constructing inscriptions for rhetorical purposes. Thus, rather than being required to produce pre-specified procedures, inscriptions, and solutions, students were free to decide how they wanted to make their case. We analyzed graphicacy in the context of the solutions that had required students' rhetoric strategies in a double sense. At first, they had to convince their group members of their own understandings and proposals for doing the task. Then, as a group, they were asked to produce a claim and provide supportive evidence. This focus on the rhetorical aspects of graphicacy was evident in the transcribed episodes and rationales provided for using graphs. From this perspective, the classroom use of story problems has taken on new dimensions. Previously we thought that these problems were (a) occasions in which students could practice their skills of decoding complex information and the application of mathematical knowledge; and (b) ecologically valid measures of students' mathematical practices during field work. Now we see these problems in a different light.

The present investigation shows that fieldwork and story problems gave rise to different *interpretative* and *productive* graphicacy; the fieldwork situation also became a resource in the story problem situation. However, both situations provided occasions for experiencing and seeing intelligible graphicacy. In this sense, we feel that both are valuable situations, provided that the meaning of context is changed. Rather than referring to the situation described by the word problem's cover story (con-text), a word problem should be termed contextual if it gives rise to intelligible mathematical practices embedded in a range of other mathematical and other scientific practices within a classroom community. Word and story problems are then both instructional attempts in which students can draw on previous experience (also provided in the classroom) and a fertile ground, in which rich and problematic discussions can occur. Here, graphicacy is not treated as something in individuals' minds, but as a witnessable social event with associated collective products. In this sense, the word problem was contextual; but the teachers might have wanted to spend more time with the entire class to discuss the various forms of graphicacy that students employed. We believe in the central importance of students' autonomy and the situations that occasion learning. The relationship between perceived autonomy and achievement in this part of the course supports this intuition. Although much more constraining than their field experience, the task setting of the story problem still allowed students to maintain a sense of autonomy; they constructed responses on the basis of their rhetorical power rather than in terms of an external standard. Given the important role of students' field experience to the inscriptional practices in the word problem situation, we recommend an increased use of such field or laboratory experience, in addition to suitably chosen word problems. Students can then draw on lived-experience as a resource in other activities that allow them to participate in and exhibit appropriate forms of graphicacy. The word problems are then contextual, because they are part of the observable practices in a consensual culture of science.

Stripping and rebuilding context

Graphicacy in our studies can be understood partially in terms of infinite chains of transformation practices (see chapter 4). In one direction, the transformations have more local and material character (e.g., when students float soil samples or produce a soil sample with a soil corer); in the other, the transformations produce inscriptions that are characterized by reduction and increasing universality (e.g., transforming a table into a plot, reducing a number of data points to one trend line). The students turned their activities of talking about their ecozone, framing a question, collecting measurement instruments, walking to the field site, and the contingent decisions into 'scribbles' in a notebook. They transformed these into more convincing inscriptions such as ordered tables and graphs. The strength of both tables and graphs derives from their ability to exploit the two dimensions of the paper surface simultaneously, although in different ways. The tables were still close to the original measurement. In their discreteness, they literally stood for the individual measurements, now ordered according to one or more factors. The next step consisted of simultaneously representing two measurements in one point. A number of data pairs were then further reduced to a 'trend', represented by a line (see the classification of graphs in chapter 2). In this reduction, there was at the same time an expansion from the discreteness of the data points to the continuity of the line. In a sense, this expansion into continuity is a move toward the continuity that we assume and often experience in everyday phenomena. A further reduction achieved with tables and graphs is the simultaneity of perception of something that took a considerable amount of time to collect. In this process of collecting and collapsing, more and more background information was eliminated to bring the phenomenon of interest to the fore. In this, the students produced cascades of inscriptions, particularly those that are deemed more convincing in the scientific community; they engaged in these practices only after they had extensive phenomenological explorations that were experientially real for them.

In interpreting the sense of the 'Lost Field Notebook', students were asked not to begin their sense making with an experience but with an inscription that summarized someone else's observational experience. At first, they sought more information about the physical context of the ecozone displayed. Upon our insistence that they do the task on the basis of the information provided, they produced descriptions such as 'a tree, shading the south-west corner of the ecozone'. This rebuilding of situation description happened here in terms of students' prior, extensive experience in natural environments, permitting students to make sense, to construct possible meanings of the trend lines they constructed. Students also interpreted the inscriptions (graphs, tables, and maps) they produced in terms of possible natural environments. The graph had to be embedded, dressed up again with possible background information, so that a plausible natural environment reemerged. That is, the students' practices in the field situation became resources (and perhaps anchors) for their mathematical practices in the context of word problems. Thus, one investigation was to measure the relationship between soil temperature and plant coverage. James and Fabricio's interpretation that the sun dries out the soil, which may lead to a decrease in the coverage, can be understood in terms of this experiment.

The students' efforts to seek relationships in their data sets were interestingly contrasted with other research. First, natural scientists produce smooth curves by fitting discrete data points, each of which is associated with some degree of random error. In the fitting procedure, these scientists produce curves that are perceived as eidetic images, rather than mental images, and thus represent the transcendental essence of objects-in-experience. The seven groups who plotted the data and constructed best-fit curves engaged in a very similar practice to induce ideal relationships from defective data. On the other hand, the groups who based their negative results on verbal reasoning only argued that the provided data did not lend themselves to support an ideal relationship. Most of these groups hypothesized the existence of a range of variables whose interplay determined the plant density. It is apparent that these students exhibited very sophisticated reasoning. In terms of the scientific ideology, however, their argument would have received increased support had they made use of an inscription.

The move from a text (word problem) to a more powerful inscription, such as a data table, graph, comparison of means, or equation, is quite complex. To make the required transformation, the various inscriptions have to be seen as objects in themselves that can be transformed into one another—in chapters 7 and 8 we investigate the work involved in such transformations. We show here, however, that the students were only beginning to increase their use of more universal inscriptions, and the need for producing the more convincing universal graphs was not yet ubiquitous in the classrooms. Furthermore, the inscriptions they produced indexed specific phenomena in specific settings.

From a discourse perspective, students' activities can be understood as that of the production of text. During the field experience, their text took on primary quality based on some non-textual experience, whereas the texts they produced in response to our 'Lost Field Problem' task were secondary. Much as in hermeneutics, secondary texts elaborate the meaning of the primary texts, they become con-text. Building the first type of texts appeared to be more sensible to students' worlds, because they could do so from and on the basis of their current knowledge and common sense. The story problem already presumed that students (a) were competent in reading and understanding stories and (b) could make sense. Thus, in addition to the explicit story context, there was another text (common sense, con-text), implicit and taken for granted by the authors of the story. The students' graphicacy in the different settings can thus be understood in terms of a broader literacy. In their field experience, they produced their texts more easily and without asking for more information, unlike in the story problem. We can conceptualize the level of context in terms of the ease with which individuals produce con-texts; this ease, however, depends on the relationship between the individuals and the text. Thus, 'contextual word problem' cannot be an ontologically fixed category to classify school tasks but is determined by the extent to which individuals are able to participate in practices, that is, produce additional, elaborating texts. In the present investigation, students brought their experience to build a context in which the distribution of light and plant density would make sense, or in which the best-fit curves would have a correspondence to some real world phenomenon.

4   Photographs in biology texts

In the previous chapters, graphicacy often involved graphs. However, despite the similarity of the words, graphicacy is not limited to graphs. Graphicacy refers to the practices of employing, reading, and interpreting all forms of inscriptions. In fact, graphs appear with a much lower frequency in high school textbooks than, for example, photographs, which dominate school science textbooks. It is surprising then that the function of photographs and their relation to captions and texts have not been the topic of much research work from practice perspectives. In this chapter, we investigate the function and structure of photographs in high school science textbooks. Our motivating research question is, 'What can students learn from textbooks when they study photographs?' To answer this and several subordinate questions, we selected and analyzed four Brazilian biology textbooks. We focus on the use of photographs and the relation between them, various types of texts, and the subject matter presented. Our analysis reveals that the structural elements of text, caption, and photographs and the relations between them differ across the textbooks and at times even within the same book. This, of course, influences readers' interpretations of the photographs and changes their role in the text.

Reading photographs

Photographs constitute a major aspect of high school science texts: there are about seventeen photographs on every twenty pages of high school biology textbooks. It is surprising then that a photograph (like a word) on its own does not mean anything (see Introduction); it is only through recurrent use in similar situations that the relation of a word to other words, a photograph to other photographs and words are established. For example, one might ask, 'What is the content of the photograph in figure 4.1?', which was taken from a Brazilian high school biology textbook. 'What is its meaning?' There are some cows in the foreground, two trees and a fence further back. Then there is a field or meadow

Fig. 538. Two distinct biomes border on
each other: forest and savanna. The latter
is a variety of field. The dividing line between both biomes is a band of higher
density of vegetation, which identifies the
ecotone.

*Figure 4.1. Example of a photograph and caption from a Brazilian high school biology
textbook in the context of teaching the concept 'ecotone'. The photograph not only provides evidence or exemplifies the boundary between two biomes but also teaches how to
identify the boundary between the two types of biomes. (Reproduced with permission of
the copyright holder.)*

before an assembly of trees, which may be seen as a 'forest'. So what does it
mean? To find an answer, we have to, as shown in chapter 3, seek some *con*text.
That is, we have to seek recourse to the text from which the photograph was
culled. The caption to the photograph talks about there being distinct biomes
('The dividing line is a band with major vegetation that defines an ecotone').[1]
Knowing this, we can now return to the photograph and attempt to discover dis-
tinctness that would delimit the different biomes that we are to find. Further
reading of the caption then tells us something about changes to greater density.
The caption also talks about forest and savanna, the latter being a kind of field.

Once we find these descriptions, our gaze separates forest and field, disre-
gards the trees in the foreground, and isolates changes in plant density. What the
text has done, therefore, is not just describe what there is in the image—if it was
only a description of something self-evident, it would not have been necessary.
Rather, the text taught us what to look for and how to parse a rather dense visual
field. The text contributes to teaching us how to detect biomes, ecotones, and

how to distinguish them—though this particular photograph makes the concept of ecotone appear in a simplistic way as a clearly identifiable boundary. At the same time, the text in itself lacked something that the image provides. Here, the figure *authenticates* what the text is about, the existence of biomes and ecotones, and the borders that exist between biomes. In sum, the texts that are copresent with the photograph—that is, the co-texts—provide the pedagogy for reading the photographic image, allowing a small rather than a potentially infinite number of interpretations to be viable. A pedagogy of photographs tells us what and how photographs and the associated texts work and work together to teach a concept.

'What can students learn from textbooks when they in fact begin to study photographs?' The question is salient particularly in the context of the present photograph because of the difficulties of making distinctions between forest and savanna experienced even by scientists.[2] Thus, Bruno Latour documents in great detail an expedition in which Brazilian and French scientists attempted to decide whether the forest was taking over the savanna or whether the savanna was taking over the forest. A major problem to be resolved by the scientists he studied was just where to locate the boundary between forest and savanna; another sociological study of ecologists also showed the tremendous collaborative work that went into deciding what constitutes the boundary between forest and bush. Similar difficulties have been observed among amateur birdwatchers as they attempted to identify birds even though they had the photographs of their bird field guide directly in front of them.[3] In practice, therefore, making a distinction between forest and savanna appears much more difficult than the high school textbook leads us to believe. As science educators, we therefore question, what can and do photographs achieve when they are used in high school textbooks? What purpose do photographs serve if they cannot guarantee that students identify their equivalent in the natural world—after all, are science students not supposed to understand and be able to explain the world around them? In this chapter, we analyze the use of photographs in high school biology textbooks, what these photographs provide for meaning-making processes, and the relation between them and the subject matter presented in the text. This chapter was motivated by the goal to understand the prevalence, function, and structure of photographs in biology textbooks, that is, to better understand the pedagogy of photographs in high school science. The textbooks chosen as data sources are widely used in Brazil.

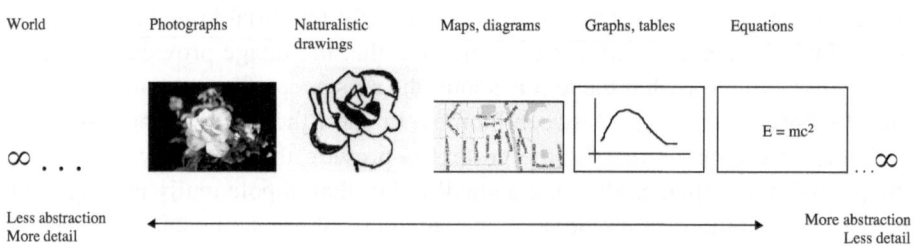

Figure 4.2. Inscriptions lie along a continuum depending on the amount of contextual detail that they carry in the background of the central object proper. However, between any two neighboring inscriptions, there is a gap in kind, which, in use, has to be bridged by graphicacy practices.

Cascade of inscriptions

Inscriptions that stand for natural phenomena usually appear first in scientific laboratories and field research sites, and—after having been cleaned, superposed, transformed—are later used in scientific publications. The more information an inscription summarizes, the more it becomes complex, resistant to deconstruction, and, therefore, powerful. However, the amount of information that can be summarized in an inscription also determines its abstractness. That is, the more information is collated together into one and the same inscription, to more contextual detail is being dropped. Thus, a photograph of a rose represents one particular rose, whereas a drawing of a rose may stand for roses more generally (figure 4.2). This claim is supported by research that shows that it is easier to classify birds using a field guide with drawings, which represent birds more generally, than using a field guide with photographs, which display birds more specifically and with considerable background detail.

From a scientific perspective, the elimination of gratuitous detail is part of the move from the particularity of one observation to the generality of a scientific claim. Therefore, photographs are placed at one end of this set of categories, presenting the background as a space continuous with our own lived experience, and, then, full of gratuitous detail. All this detail may not carry relevant information; however, it does have a function, making the photograph seem to be an extension of nature into the pages of the book, even though the effect of realism does not depend on the complete reproduction of the world. The photographs obtain their powerful role as representations of the real world through the reader's work of interpretation, the viewer's perception of the narrative and perceptual order of the document.

Whereas there is a preference for inscriptions to the right side of figure 4.2 in professional science, high school science textbooks predominantly use photographs and naturalistic drawings.[4] This is perhaps correlated with the fact that photographic and pictorial inscriptions are more likely to have an impact on individuals outside of science than graphs or equations that are often incomprehensible to students and other lay people. In the educational literature, there exist a number of reports that focus on students' use of representations together with textual information. However, despite the large range of possible instructional functions of inscriptions, research provides little evidence that inscriptions live up to their potential in print and computer media.

Most students are familiar with photographs in general; however, appropriate instructions for how to read and analyze photographs are currently not provided to them. In particular, students generally neither receive instruction in critical analysis of photographs nor are provided with opportunities for participating in the associated practices. Outside education, a small number of studies considered the role of photographs in communication. Such studies show that photographs come with apparent self-evidence. Because of the similitude with the objects represented, photographs are taken as prima-facie evidence, that is, as guarantor of truthful representation. Because photographs are automatically produced, eidetic images of the object portrayed, their constructed nature normally disappears. To understand the effect of photographs in learning, we must consider not only the way they are produced, but also the way they are received by the reader.

Photographs and texts: principles of analysis

We began this chapter by making the point that a photograph in and of itself means little; it is full of 'gratuitous' detail that allows many different ways of looking at and interpreting it. This photographic detail provides a space that is continuous with our own lived world, allowing readers to establish a link with the everyday world that surrounds them. At the same time, the photograph provides few cultural codes (e.g., a line, letter, or recognizable shapes) that could delimit its sense and meaning as intended by the author (see our Introduction, in particular the opening quote by John Berger). To control the range of possible meanings that a photograph can give rise to, authors use captions and embed this

Fig. 577. The European partridge, during the winter, shows white plumage, blending with the snow. At the end of the winter, it starts to change its plumage, and acquires a coloration that blends with the dry vegetation where it lives. This is a good example of *camouflage.*

Figure 4.3. Example of the relations between photographs and its captions. The text not only describes what is to be seen but also constitutes a lesson of how to look at and analyze the photographs. The sequence of photographs directs the reader to notice invariant features in the face of the evident variations. (Reproduced with the permission of the copyright holder.)

photograph-caption combination in still further text (main text) that together constrain the meaning a reader can make. In this section, we propose some principles for the analysis of photographs in school science textbooks.

From a semiotic perspective (see chapter 1), the photograph and its caption are two different sign assemblages or two different texts, where text refers more broadly to any entity that can be interpreted. However, caption and photograph are not independent. As captions always appear just below or next to a photograph, the two different and arbitrary sign forms are directly associated with one another. They are said to be about the same thing. Take the photographs and captions in figure 4.3 that appear in the context of a textbook treatment of 'camouflage'. In the last sentence, the caption suggests, 'This is a good example of camouflage'.

The caption articulates 'winter' and 'white plumage', which calls attention to the color of the plumage during a particular season, winter, which, *if* students are familiar with snow, is easily identified with the left-most image in the figure. (Most of our Brazilian students have never seen snow and know it only through the media but not through their lived experience.) Simply by the fact of being articulated, even if it was not named, the white color of the plumage in winter is likely to become salient. Even more so, the presence of three images inherently calls for a comparison of seasons and plumage across all photographs, and thereby makes salient the changes across the three images. That is, the text elaborates and therefore teaches how to read each image and the sequence of images (to understand change). The presence of three images calls for comparisons to identify variant and invariant perceptual properties of what the caption marks as being the same animal.

However, the photographs also (and reflexively) elaborate the caption text in the sense that they provide evidence for particular statements. The photographs validate what the text states, the whiteness of plumage and the changes the plumage undergoes; the text elaborates how to read the photographs, contributing to its sense. Here, we have two forms of texts, and two kinds of literacy involved, one verbal, one visual, each elaborating the other in their respective relations to the thing that they are about: the idea of camouflage. This 'idea' is the real referent of the word and what is to be seen as difference in the collection of photographs. As arbitrary signs, both stand in an open and yet-to-be-elaborated relation to their content, the entity that they are about. The 'this' in the statement 'this is a good example of camouflage' can be read as an indexical reference to the content of the previous sentence(s), which describes how to read the photographs, and to the content of the series of photographs, which legitimate the text.

It may be useful here to draw on the notion of intertexts, which are all the other texts readers use to make sense of and that therefore serve as a background against which he or she reads the primary text. Without the intertexts, one would not be able to make sense, for, as we showed in chapter 3, everything makes sense only against some background of other things, which constitutes the *con*text against which all situational text is seen as figure against ground. That is, graphicacy is never a literacy that stands on its own but is always tied up with other forms of literacy and lived experience. In the present situation, the caption and the photograph are about the same thing (co-thematic) and pertaining to the same activity structure (co-actional) but are of different genre (non-co-generic) texts. Texts and photographs are semiotic resources that are co-deployed (i.e., are systematically or strategically used together) and, becoming intertexts for one another, elaborate each other.

For the analysis of textbooks, we developed a scheme that articulates various semiotic resources and the nature of their relations (figure 4.4). We view all relations between the different parts of a book (main text, figure, caption, and [sidebar] text box) as involving double movements, each pair of entities mutually constituting one another and the relation. Thus, the title prepares readers for what is coming, and thereby organizes their reading. At the same time, a title is not chosen arbitrarily, but has been motivated by the content of the main text. The main text makes certain claims or seeks to explicate a concept, which there-

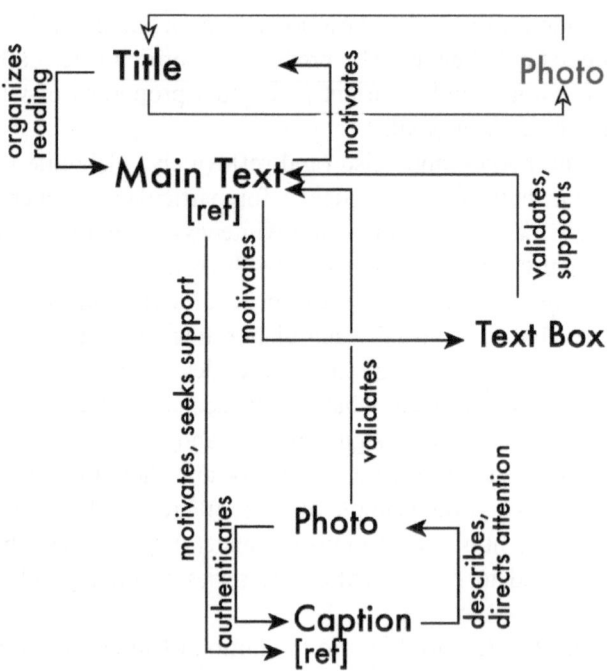

Figure 4.4. The framework developed for the analysis of inscriptions that accompany scientific texts in general and for photographs in particular shows how the different elements of a textbook page (large type) are related to one another (verbs, small type).

fore motivates the use of a particular figure. The figure in turn validates the claims made in the main text. Finally, the caption describes and teaches how to read the figure (here photograph[s]) and the figure authenticates the caption text.

Here, we are centrally concerned with the relationship between photograph (figure) and caption, and their integration with the main text. This integration is achieved not only through the co-thematic nature of figure and caption but also through an index (indexical reference) by means of which readers are referred from a particular place in the main text to figure and caption, which constitute a different genre. The caption is an essential part of the inscription that tells readers what look for in the photograph and therefore how to read and understand it. The photographs are associated with a text that explains the phenomenon. Thus, photograph and text together form the written correlate of a demonstration; they constitute a particular form of pedagogy, though our informally acquired information in another study shows that most students disattend to anything other than the main text. If this is the case, then important concepts and information should be placed in the main text, with the appropriate reference to the inscription that would help the reader to make sense of the phenomenon under scrutiny.

In our analysis of the pedagogical role of photographs in high school biology textbooks, two themes emerge: there are different functions that a photograph and its caption have with respect to the main text and there are different ways in which the photographs and the texts are structured, with implications for the interpretation of these inscriptions in the textbook. In the following sections, we describe and provide evidence for these two themes.

Functions of photographs in high school biology textbooks

In the ecology units of the four Brazilian biology textbooks, all photographs could be classified as fulfilling one of four functions (roles), which arise from the relation of photograph and its caption to the main text. However, we also include in this classification those photographs that accompanied other inscriptions, as for example, maps. Series of or pairs of photographs were considered one single inscription in this classification.

We identified four functions including decorative, illustrative, explanatory, and complementary functions.[5] These functions—and therefore our categorization—largely arise from the interpretation of caption, the text co-deployed and directly associated with each photograph. These functions also roughly define a hierarchy of increasing informational value (explaining a concept does more than simply illustrating a concept) and those with higher information value usually also do what the photographs of lesser informational value do. We exemplify and discuss each of these roles.

Decorative function

A small number of photographs are decorative. These photographs are not referenced in the main text, do not include a caption, and usually appear at the beginning of a unit, chapter, or section of text. Figure 4.5, for instance, appears on the opening page of a section on 'energy and matter in the biosphere'. This photograph does not include a caption; there is no reference from the opening of the main text to the photograph. How the photograph functions in relation to other texts deployed (its intertextuality) requires analysis and does not 'jump out' at the non-initiate. At the outset, it is a colorful plate from which relevant figure and ground have to be separated. Prior exposure to cultural categories allows readers of a certain age—a one- or two-year old may not perceptually differentiate what an adult sees as leaf or caterpillar—to identify a caterpillar on a leaf.

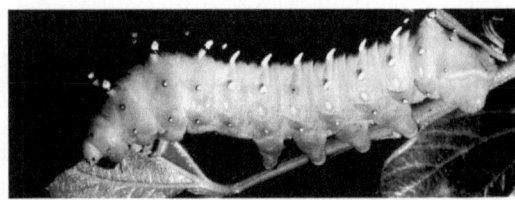

Figure 4.5. Example of a decorative photograph; there is no caption that accompanies it and the main text does not refer to it. There is therefore no other indication of a function than that the photograph decorated the chapter beginning. (Reproduced with the permission of the copyright holder.)

That is, in the absence of a text inscribed in the book with reference to the photograph, the reader has to bring existing understanding as the intertext in reference to which the photograph becomes salient figure. What is the role of this photograph at this place in the book? What can a student learn by looking at or analyzing (studying) the photograph?

A photograph can be viewed in many different ways (see opening quote in the Introduction). To understand what *this* photograph is intended to show in *this* place, a reader may search for clues in nearby texts, such as the title of the unit. Assuming that the text is not only co-deployed but also co-thematic with the photograph, a reader seeks to relate individual words 'energy', 'matter', or 'biosphere' to the photograph. A somewhat initiated reader may see the caterpillar 'nibbling away on' or 'eating' the leaf; but we insist that 'nibbling' as a process is not available to readers, it has to be inferred based on extra-textual experience. However, not until readers know the relationship between 'eating' and 'energy household' of animals can they establish a connection with (one part of) the unit title. At the same time, the leaf has to be understood as matter rather than as an organism, and both the caterpillar and the leaf have to be seen as aspects of 'biosphere' before the relation of this photograph to the unit title can be established. Students, however, are not likely to bring this understanding necessary for establishing these relationships between unit title and photograph. In fact, the purpose of the unit is to develop the understanding necessary to deconstruct the relationship between photograph and title.

This initial analysis shows how, for the initiate reader, unit title and photograph can be seen in a mutually constitutive relation expressed in figure 4.4. The word 'energy' makes a reading of 'caterpillar eating leaf' a reasonable reading of the photograph, which, in turn, establishes a concrete instance of the relationship between biosphere and 'matter and energy', concepts usually introduced in

Fig. 4.3 Photograph of plants of aguapé in blossom.

Figure 4.6. An example of an illustrative photograph. The photograph gives an example of a particular object, a token illustrating the type (aguapé) evoked elsewhere in the textbook. (Reproduced with the permission of the copyright holder.)

the physical sciences. However, because students do not (usually) bring the interpretive resources required for the type of analysis provided and because of the lack of a text that could guide students in their analysis of the image, we categorized such photographs as decorative. They introduce color, may provide for certain aesthetics, but lack informational function for the individual who does not already know what the subsequent text is intended to teach.

Illustrative function

Photographs included in this category include a caption that names or describes what the reader is to see in the photograph but the caption does not provide additional information to the main text. Such photograph-caption ensembles constitute a visual resource for the reader in the sense that a concrete specimen of a class or concept is depicted. Take, for example, the depiction of an aguapé plant and the caption that was associated with it (e.g., figure 4.6). This photograph gives the reader a visual representation of the species articulated in the main text, the aguapé plant, but this is not an essential piece of information for the reader relative to the subject matter treated in the text. In the present case, the subject matter is the introduction of certain species in biomes, exemplified by the introduction of aguapé in hot regions. 'Aguapé' and 'hot regions' are special instances, concrete realizations of the more general concepts of plant and biome.

The photograph illustrates the particular plant but does not show 'introduction' that causes changes in the ecosystem. To show the effect of 'introduction' of a plant, a number of photographs are required that show some difference that can be noted as a difference before it can function as 'information'. That is, if there is not a difference that makes a difference, we cannot speak of information at all. Therefore, the very concept taught in the text is absent from the photograph: it does not exist as information in the image. The visual information possibly provided does not alter the understanding of the subject matter, that is, the photograph does not show the phenomenon treated in the text, but provides a visual illustration of a plant that was only referred to in the text as an example of a species which introduction caused changes in the ecosystem. The reader still is able to understand the concept of ecological disequilibrium treated in the text without the information provided by this photograph and the caption.

There were several cases of illustrative photographs that were not associated with a (part of a) caption. Such photographs, a special case of photographs without caption, appeared together with 'maps', the dominant aspect of the inscription (figure 4.7). Here, several photographs were co-deployed with the map but were not described or explained in the caption. One might therefore think that the photographs are decorative, especially because the caption of the inscription is related with the map. However, there is an important link between photographs and map: the color scheme of the legend relates photographs, presenting single (paradigmatic) instances of different landscapes, and regions. If map and photographs are interpreted as being co-thematic, by virtue of appearing in the same plate, the different genres can be read as linked via the concept of biomes: 'distribution of different *biomes*' and concrete instances of individual biomes. In this situation, there is one photograph for each biome but, in the presence of six images, a contrast is provided between what may be prototypical examples for each biome. The presence of only one example does not allow students to learn what characterizes each biome or more poignantly, how to distinguish one biome from another in more problematic cases near the border of the category.[6] But the presence of six prototypes, given learners attend to appropriate aspects of the landscapes depicted, may allow the recognition of some global distinctions between these biomes. Nevertheless, some of the very features that distinguish these biomes, the amount of water available, temperature, and other physical and biological information, are not accessible by students through the analysis of the photographs.

Explanatory function

This category includes photographs with captions that provide an explanation of or a classification of what is represented in the photographs. The captions do not only name the object or phenomenon in the photograph, but also add informa-

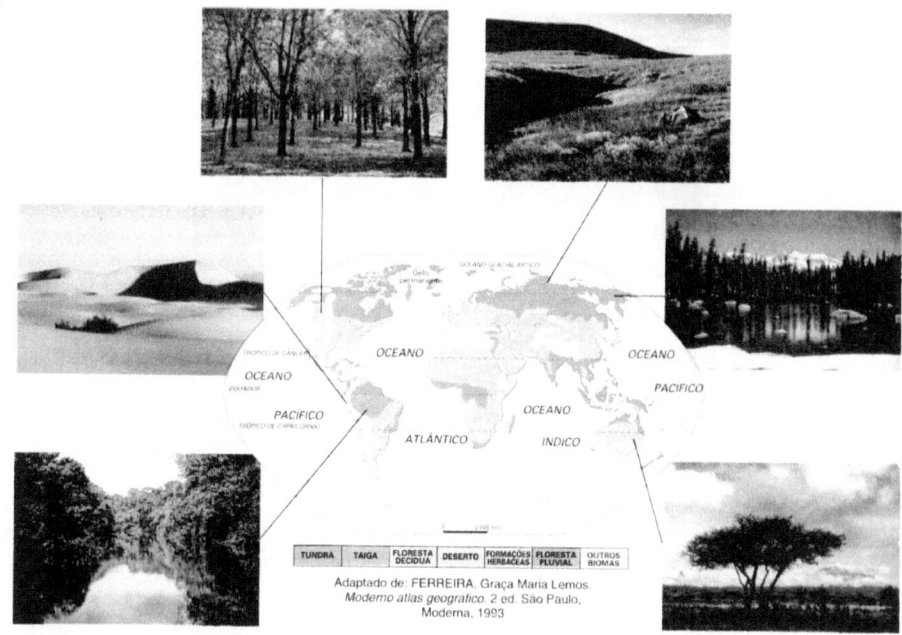

Fig. 7 The large biomes of the Earth.

Figure 4.7. An example of a special case of illustrative photographs, each biome identified in the map exemplified by one member of the category. (Reproduced with the permission of the copyright holder.)

tion about this object or phenomenon. Take the example of figure 4.8. In the first part of the caption we can read 'Aspect of a forest'. With this information, readers are guided in what to look for in the photograph, a forest. The sentence is part of a story that shapes how we look and what we see; the story gives our gaze a direction and purpose, which shapes the perceptual field of which we become aware. That is, what we see are not just a group of trees along a river but part of a larger whole. This information provided by the caption is important in helping the reader to make sense of what can be seen in the photograph, however, this information is not enough to guide the reader to establish relations between the photograph and the subject matter treated in the main text.

The index 'Fig. 84.1' presented in the main text and replicated in the caption allows the reader to connect figure and text. However the reader is not able without further information to appropriately relate the 'forest' in the photograph

Fig. 84.1- Aspect of a forest: climax community.

Figure 4.8. Example of an explanatory photograph. The expression 'climax community' provides a frame that allows the reader to establish a connection between the figure and the main text. (Reproduced with the permission of the copyright holder.)

with the concept of 'ecological successions' that is the corresponding topic of the main text. Thus, if this were the only information provided in the caption, the photograph would function as an illustration of a forest, because somewhere in the main text the forest was mentioned. It is the second part of the caption that provides the information necessary to interpret the forest in the photograph as 'something else', which allows the reader to explicitly relate the figure and the text. These two words, 'climax' and 'community', represent an entire different perspective in the way in which the reader contextualizes the photograph and relates it to the main text.

The photograph not only represents a forest, but also is *marked* as an example of a climax community. Textual marks are not neutral but invite making salient some things to the exclusion of all the others that could be made salient. That is, marked terms encourage readers to associate the characteristics of a climax community described in the main text with what they see in the photograph. In this sense, this caption not only classifies the forest as a climax community, but also provides an explanation about how to interpret and relate the photograph with the main text.

At the same time, because this is a single photograph, the concept of succession is not available to readers, which would require several photographs showing the same physical location but with varying cover corresponding to

Fig. 86.6- Linofrino, an example of an abyssal fish, about 5 cm long, that lives at a depth of 1400 m in the ocean. The abyssal fishes usually are small and dart-like and have sensitive eyes.

Figure 4.9. Example of complementary plate: the caption makes factual statements not available in the main text. (Reproduced with the permission of the copyright holder.)

varying stages in the ecological succession of the area. Similarly, the single photograph does not allow the initiate reader to learn how to distinguish climax forest from non-climax forest, or between the climax forest for different forms of successions such as those that end in maple-beech forest (Northeastern US, Eastern Canada) or those that end in coniferous forests (Canadian shield, Newfoundland). Both types of forest are examples of climax forest but are very different in the way that they appear to the eye.

Complementary function

Photographs in this category are associated with captions that add new information about the subject matter treated in the main text. This information is not only new, but it is also an important information, never mentioned before in the main text, and that helps readers to further understand the biological concept that is being taught. Figure 4.9, for example, presents two fishes against a black background. The title of the section of the text where this figure is inserted in is 'Influence of light in the marine ambient', and in the main text we can read about the distribution of species in the ocean according to the presence or absence of light. In the last paragraph the text presents some characteristics of fishes and other animals that live in the abyssal zone. Then the text refers the reader to the figure.

The text begins by providing a name and articulating it as an example of an 'abyssal fish'. Inherently, the statement 'linofrino, an example of an abyssal fish' requires the cultural competence of associating the name with the image, even though there is no specific index linking the name with the fish—parents reading to their preschool children might place their finger on the image and say, 'linofrino'. The remainder of the caption provides propositions with content not made available in the main text, and therefore constitutes new and relevant content. We therefore classify this photograph-caption ensemble as complementary.

The caption in this case provides information about what can be seen in the photograph, that is, characteristics of the abyssal fishes represented. The caption also adds *new* information, not directly related to the two fishes, but, rather, associated with the concept of abyssal fish treated in the text. Therefore, this plate constitutes a 'complement' to the main text. The complementary photograph-and-caption combination thus presupposes continuity in the reading process, as readers iterate their reading between main text and plate. They are able to make sense of the concept presented only through the reading of all the information contained in these three elements. *If* the information in the caption is new and important, we therefore have to ask why this information is in the caption instead of in the main text—unless students and teachers regard photographs and their captions as integral parts of the 'material to be studied'.

Structures of co-deploying photographs and texts

Our analysis reveals considerable variation between and within textbooks in how the co-deployment of photographs (figure 4.4) and texts is structured. The structural elements and the relation between them are diverse among the selected textbooks and even within the same book. The variations include, for example, *where* the indexical reference to a photograph occurs in the main text, the distribution and arrangement of photographs on the page, and the co-deployment of multiple photographs for teaching a particular concept. These structural elements, undoubtedly, provide different semiotic resources for integrating co-deployed and co-thematic but non-co-generic text.

Indexical reference

Photographs represent a different genre than text. They are two-dimensional arrangements of colored areas, which, because of our prior experience in a three-dimensional world, can be decoded to provide additional information about depth. Color, areas covering other areas, relative size of known objects and so forth provide resources for reading that are deployed as the eyes scan the image according to the reader's preference. Verbal texts, on the other hand, are linear. Because of the different requirements for reading verbal text and images, the latter cannot be placed at the point in the text that is directly pertinent (co-thematic).[6] The link between the text (word, sentence, or paragraph) and the photograph that appears somewhere else on the same or different page is established via an indexical reference usually as a string of letters and numbers in the form 'figure 4.1.2', 'fig. 2', or 'see fig. 3.4', which is also found in the caption of the figure that the index is designed to direct the reader. Here, the indexical function is achieved by duplicating a string in the main text and in the caption. Whenever the string appears in the main text, the reader is referred to the photograph/caption that features the same (co-generic) string. That is, the relationship and 'placement' of a photograph with respect to the dominant text is achieved by means of a string that appears twice but in different locations on the page or in the book. The role of the string is salient when we consider that a similar relationship does not suggest how to link caption and photograph. In this case, physical proximity is used to suggest that the text directly bears on something in the image. How this bearing might be achieved still remains undetermined at this point. (We discuss the nature of this relationship and how the reader enacts it in the next section.)

Two textbooks consistently used the same way of referencing photographs and captions either placing the indexical reference at the end of a paragraph in which the co-thematic concept appeared or not using an indexical reference at all. The two other books each employed three different ways in placing the indexical reference in the main text. Thus, the indexical reference was placed either at the end of the paragraph or directly with the co-thematic word or sentence or was absent altogether.

When the indexical reference is placed immediately after the word or after/within the sentence that is co-thematic with the photograph and caption, a *direct* link is established between what are on the surface different (because non-co-generic) representations (figure 4.4). On the other hand, if the indexical reference is placed at the end of a paragraph where there are potentially multiple concepts presented, the link is no longer direct. One may consider the index 'misplaced', because the photograph and caption are not evoked simultaneously with the verbal texts. There is the potential that misplaced indexical reference in books interferes with sense-making processes in ways similar to misplaced gestural indexical reference that make it difficult to learn from lectures. Finally, when there is no indexical reference at all, it is totally up to the reader to see whether there is any relation at all between a photograph/caption and the main text on the same particular page.

a.

Factors of Ecological Disequilibrium
Changes in the structure of ecosystems

Deforestation

One of the most important ecological
problems today is the destruction of forests, as had occurred with the Atlantic
Forest in Brazil. Today less than 10% of
this forest type remains compared to the
period of colonization. Each year, the
world loses forest areas; forests are cut or
burned, leading to serious soil damage
and causing atmospheric pollution. Furthermore, many species become extinct,
thereby decreasing 'global biodiversity',
as scientists call the large variety of living forms produced by biological evolution. (Fig 4.1)

b.

Figure 4.1 Deforestation is a common way of
damaging terrestrial ecosystems. (A) Photograph of burned area in the Amazonian Forest,
used to create pasture areas for livestock farm-
ing. The fire kills the microorganisms that fer-
tilize the soil, and the rains wash the nutrients
away since the vegetal coverage was destroyed.
(B) Photograph of regions of soil erosion pro-
voked by the elimination of the forests.

*Figure 4.10. Example of an inscription with a 'misplaced' indexical reference. a. Main
text. b. Photograph and caption referred to at the end of the corresponding paragraph in
the main text. (Reproduced with the permission of the copyright holder.)*

Figure 4.10 exemplifies a 'misplaced' indexical reference. In this situation,
because the index is placed in the end of the paragraph, the reader may associate
the photograph more spontaneously with the last phrase or statement—
particularly in those textbooks where the indexical referencing varies. The photographs represent (a) a burned area and (b) an area of erosion of the soil. Although the main text mentions burning and cutting the trees as ways of causing
deforestation, the index to the photograph is physically away from the specific
phrase where deforestation is mentioned.

To associate the photographs with the main text, the reader needs to go back
to the middle of the paragraph and find the specific phrase that refers to deforestation. Then, the reader must go back and forth in his or her attempt to read the

Figure 10 Quantity of water and richness of life are interdependent.

Figure 4.11. Example of a photograph without any indexical reference in the main text. (Reproduced with the permission of the copyright holder.)

text. This reading requires the reader to work from the text and the photographs, at the same time 'reading' and 'seeing' to make sense of the biological concept presented.

Figure 4.11 presents an example of an inscription with no index in the main text. This inscription presents two photographs: the photograph to the left shows a river with abundant vegetation in both its banks, and the photograph to the right shows a desert. The caption reads, 'The amount of water and the richness of life are interdependent'. Because the main text does not contain an indexical reference, we can only try to establish a relation between text and photographs, the relationship of the photograph to the main text can only be subsequent to reading the entire main text section and the photograph with caption.

When the photographs appear alongside one another, the composition highlights the importance of water for the existence of life. Thus, through a comparison of both photographs, the reader is supposed to associate life and water in the way intimated by the caption. However, this is just one way to interpret this inscription, and many other interpretations can also emerge since there are no explicit directions or enough information in the main text or caption to help the reader to make sense of this.

The situation becomes even more difficult when the photograph is physically placed far away (several pages) from the corresponding text. In this situation, readers will find themselves completely 'lost in the book', since they will not find any direct association between the text and the figure, because of the

Fig. 581. At left, the viceroy butterfly. At right, the monarch butterfly, which has repugnant taste for the birds. Because of the similarity between them, many birds reject the first one, which benefits itself from mimicry. (Reproduced with the permission of the copyright holder.)

Figure 4.12. Example of photographs arranged in pair, which allows the reader to look for variant and invariant properties.

absence of the indexical reference. Furthermore, the reader will have difficulty to manage the book pages to associate the figure with the text, because of the disposition of the inscription many pages after the one in which the text was placed. At this point, the figures even though associated with captions, may serve decorative rather than higher functions in the text.

Single and multiple photographs

The arrangement of the visual document within the text mediates our ability to see the phenomenon represented in the photograph, that is, part of our interpretation of the photograph depends on the way in which the figure is organized, and how the photograph relates to other photographs. One way in which a photograph can be related to others is as part of a pair or of a series. Photographs arranged in series allow readers to progressively focus their attention on the concept examined by the text. Consider for instance figure 4.12. At a first glance, the reader may see these two photographs as presenting the same butterfly, due to the enormous similarity between the species represented in both photographs. However, the caption cautions us that the photograph at right presents one species of butterfly, and that, actually, the butterfly in the photograph at left only seems to be the same species as the earliest, what constitutes the phenomenon called mimicry.

The authentication of the phenomenon of mimicry presented in the main text is possibly due to the arrangement of the photographs in a pair, which allows the significant differences to become evident through the process of com-

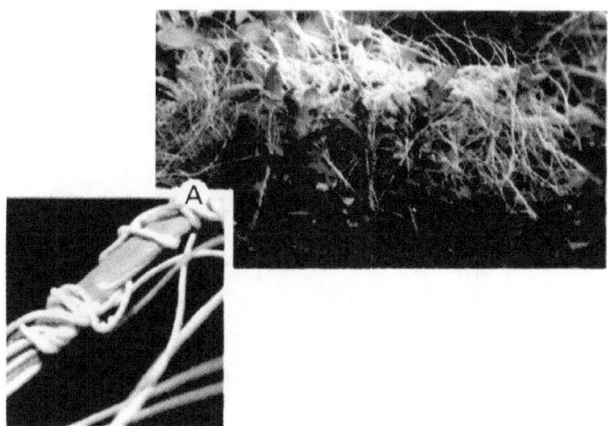

Fig. 3.10 (A) Photograph of a Hibiscus plant covered by the cipó-chumbo. In the detail, in higher magnification, the relation between the parasite and the stem of the host plant.

Figure 4.13. Example of multiple photographs, one being constituted by the text as presenting the detail of the other. (Reproduced with the permission of the copyright holder.)

parison. Nevertheless, a comparison between the two photographs is not enough to give the images their intended sense. The caption is also necessary to guide the reader to look for the differences—instead of the similarities that are more evident in this case—between the two photographs, to recognize the phenomenon of mimicry. Similarly, in a series of photographs, as for example in figure 4.4, the process of authentication of the phenomenon presented in the text depends on the reader's perception of the differences between the photographs. In making the photographs part of a series, the uncertainty about the meaning is reduced, and the reader, then, is able to eliminate everything that does not change, in a process that progressively highlights what there is to look at and make sense of in this figure.

Another way in which a photograph can be related to others is when it presents the same object as another photograph, but in a different way, which allows both photographs to become complementary to each other (figure 4.13). The first photograph shows the plant in a broader view, while the second photograph focuses on a specific part of the same plant. Together, the two photographs allow the reader to identify the plant in the way in which it could appear in nature, and, at the same time, pay attention to the specific detail relevant to

Fig. 83.1- Epiphyte plant

Figure 4.14. Example of photograph that carries too much visual information and it is unclear which of the many plant is the 'epiphyte plant'. (Reproduced with the permission of the copyright holder.)

the concept presented by the main text. Both photographs, therefore, function as complementary to one another, the second photograph becoming 'part' of the first one, as a detail in higher magnitude, that provides the reader with a better visualization of the phenomenon treated in the main text. Multiple photographs, therefore, allow the reader to make external comparisons and therefore visualize the phenomenon presented by the main text. A single photograph, however, can only provide internal comparisons, leading the reader to find the relevant details in the photograph on his or her own.

The background of the photograph

In the process of interpreting single photographs, the background becomes an important resource to help the reader to distinguish the relevant object or phenomenon depicted in the photograph. Explicit directions in the caption and other indications, as for example, letters or arrows added over the photograph itself, are also important resources that guide the reader's attention to the 'right' detail. Thus, figure 4.14 fails to demonstrate the object that it should represent according to the caption. The caption reads 'Epiphyte plant', but there are many different plants without distinction that allows readers to identify the epiphyte plant. (We show in chapter 6 that in fact students make salient different aspects of this image.) Even though there is a tree placed in the center of the photograph, which

Fig. 83.1- Epiphyte plant

Figure 4.15. In this photograph it is possible to identity the epiphyte plants, even though it still present other plants in the background. (Reproduced with the permission of the copyright holder.)

may draw the reader's attention, it is not possible to identify the epiphyte plant, unless the reader already knows what to look for. That is, the reader has to know what an epiphyte plant looks like in order to find it in this photograph. The difficulty, in this case, is related to the *framing* of photographs, that is, the process by which the reader narrows the perceptual field to make the relevant figure salient while pushing irrelevant elements as much as possible into the background. This may involve gestalt switches, which we sometimes announce to be seeking when we say things like, 'I have to try to look at this in another way'.

The aim in framing photographs therefore has to be making sure that it contains the least information possible, for fear of confusing the meaning. The details in the background seem to carry no relevant information at all, despite their function of making the photograph more 'natural', because it can be perceived as a depiction of a particular piece of nature. However, the effect of realism does not depend on the complete reproduction of the world, but on the viewer's perception of the narrative and perceptual order. Therefore, it could be more appropriate, at least in certain situations, to present an object against a neutral background, even if it compromises the 'reality' of the photograph as a depiction of the real world. Compare figures 4.14 and 4.15, both of which represent an epi-

Fig. 78.2- The dead organic matter is decomposed by bacteria and fungus, as we can see in this photograph of a leaf laying down in the floor for months, and in final process of decomposition.

Figure 4.16. The black background highlights the object represented in the photograph. (Reproduced with the permission of the copyright holder.)

phyte plant. The relevant element in this photograph—the epiphyte plant—is clearly distinct from the background, even though the photograph still presents other plants. The epiphyte plant is not only in the center of the photograph but also the only object that the reader can clearly distinguish. This figure is framed to show just this particular plant, and all other objects are out of focus, thereby becoming part of the irrelevant details in the background. That is, the responsibility for arriving at the intended interpretation of a photograph does not only lie with readers, but also with the authors who have a variety of means to aid in the process. We already made a similar comment while discussing students' problems with reading velocity from distance-time graphs and argued that it is poor practice to hide, cover up, or fail to make more salient the feature to be presented.

Sometimes a completely black background is a better alternative for highlighting the phenomenon or object in the photograph (figure 4.16). The reader is immediately directed to whatever is shown against the black background that is easily identifiable as irrelevant. Thus, the arrangement of the photographs in the books, as well as the intrinsic characteristics of the photographs themselves, have an impact on the process of interpreting the figures, and consequently, in reader's ability to relate the photograph with caption and main text.

The pedagogy of photographs

We live in a visual culture and visual representations pervade our lives. This is especially true in the sciences that historically are associated with the emergence of representational practices. In this chapter, we show that photographs can play different and important roles in the texts and that the photographic images and captions are often inappropriately referenced from the main text. That is, the photographs and captions almost function in a stand-alone mode. We began this study of photographs because our informal survey of students at all levels of education showed that they hardly ever attend to the photographs despite their abundance in the textbooks. This tendency, of course, has to be seen in the context of school science and university lessons that primarily focus on language (and mathematical formulae) as a carrier of scientific knowledge. We hypothesize that the true potential of photographs as pedagogical resources has not yet been achieved. The analysis of the photographs reveals that the structural elements and the relations between them are diverse among the textbooks and even within the same textbook. Although the main function of the caption is to help understand and interpret the photograph, the 'information' in the caption can vary, alternating the relations between the whole inscription and the other structural elements of the text. This alteration also changes the role of the inscription in the text. We identify four functions of photographs, decorative, illustrative, explanatory, and complementary.

The differences in the information provided by the caption not just influence readers' interpretations of the photograph and therefore what they can learn from them but also change the role of the inscriptions in the text. For example, we can infer from a photograph without a caption that, because there is no direct association to the text, its role is a 'decorative' one, and so the book is more 'beautiful', 'colored', and 'full-illustrated'. A caption that just identifies what is represented by the photograph gives us a clear idea that this photograph is playing an illustrative role. In these cases it is not even necessary to have the picture; it is a supplement to the text, adding details or specificity, or illustration of it. But a supplement is part of and not part of the text at the same time: it seems to be adding something to what is complete in itself, and the addition is thus implicitly a correction, sometimes to the point of a recantation.

The indexical reference in the main text also varies in the four textbooks and even in the same book. Two of the four textbooks analyzed present patterns related to the index; one of them always presents the index in the end of the paragraph, and the other one does not have an index at all. The other two books present three different ways to place the index in the main text: in the end of the paragraph; in the middle of the text, just after the phrase or the specific word related to the inscription, or with no index at all. The way in which the index was written also varies in these two textbooks, that is, '(figure 4.1)', '(see figure 1)', '. . . as we can see in figure 4.1', and so forth. Our analysis shows that these differences in the indexical reference can change or reinforce the role of the inscriptions in the text and that the relations can be established between the photographs and the other structural elements of the text.

Textbook authors use many strategies for relating photographs to titles, main text, indexical reference, and captions. The use of individual photographs to illustrate a concept appears to be limited because variation between exemplars is not expressed but necessary to understand just what a photograph depicts. Some of these ways appear to interfere with sense making, whereas others potentially offer support in the construction of meaning. Evidently, misplaced indices, that is, indices that 'point the reader in the wrong direction', disrupt the process of sense making. How do readers use the various resources highlighted in this chapter? In chapter 6, we report on a series of interviews using some of the photographs in this chapter, with and without caption and text, which had been conducted in Brazil.

5   Graphicacy in lectures

Human communication is not limited to textbooks. Teachers, professors, public speakers, television announcers and others make use of a range of meaning-making resources other than language—prosody, gestures, body movements, and the like—that their audiences use to make sense. If such talk occurs over and about inscriptions, graphicacy takes on new dimensions. First, from the speaker's side, it includes the articulation of semiotic (meaning-making) resources that highlight features of the inscription; second, from the audience perspective, graphicacy includes the knowledgeability involved in making use of these other features to understand the inscription. In this chapter, we take a look at photographs in the context of presentations where the speaker stands next to the projected photographic slides. We are concerned here with the semiotic resources that critical graphicacy must take into account.

In the previous chapter, we showed that a number of structural features of photographs in Brazilian high-school biology textbooks were shown to pose potential problems for making sense from the photographs. Because of the abundance of contextual detail, photographs lend themselves to being perceptually structured and therefore interpreted in different ways so that what exactly is important to the scientific phenomenon to be taught is often not discernible by the reader. We showed that neither caption nor main text were able to constrain this tendency of photographs to proliferate meanings rather than to constrain the number of ways readers can make meaning when they read a particular scientific text.

When presenters or teachers use photographs during a lecture, we may expect them to deploy semiotic resources not available to the reader of a text. For example, under certain circumstances, we expect lecturers to use gestures, pointing sticks, or lasers to point to a particular spot in the photograph; linguistically, this pointing constitutes a *deictic* gesture. Furthermore, lecturers can also use *iconic* gestures, which get their name from the perceptual resemblance of gestures with the phenomena they depict—following the shores of a lake on an aer-

ial photograph or using two hands to illustrate the confluence of two creeks constitute iconic gestures. That is, learning from photographs in the context of lectures can be expected to be a very different kind of work than learning from
photographs in textbooks.

The purpose of the present chapter is to investigate the semiotic resources
that lecturers make available to their audiences; such resources can assist in better understanding just what photographs are intended to express. We look at
cases in which speakers talked about scientific topics in the presence of photographs. We focus on 'lectures', that is, situations in which one speaker talks
about scientific issues while addressing a larger, mostly listening audience. We
analyze talk, gestures and relative position of speakers with respect to audience
and to photographs, with a particular interest in how a speaker's words, gestures,
and body orientations assist in making salient photographic detail pertaining to
the scientific issues at hand.

Semiotics of photographs in lectures

High school biology textbooks heavily draw on photographs; one might be
tempted to think that this means a lot of information is made available by visual
means. However, as we saw in the previous chapter, photographs alone contain
too much undifferentiated information and therefore mean little; they are full of
'gratuitous' detail that allows many different ways of looking at and interpreting
them. On the one hand, this photographic detail provides a space that is continuous with the lived world, allowing readers to establish a link with the everyday
world that surrounds them. On the other hand, this detail provides few cultural
codes that could delimit the photograph's sense and meaning as intended by the
author. This means that users of photographs may have to invent codes, which
are the result of their perceptual structuring. However, the range of possible
ways of structuring can be constrained. Thus, to control the range of possible
meanings that a photograph can be integrated into, authors use captions and embed the photograph/caption combination in still further text (main text) that together constrain the sense a reader can make. That is, the text constitutes a specification of order of coherence from which the rational visibility of the scientific
object emerges. Therefore, when there are only photographs and texts, such as
the science textbooks or scientific journals in previous chapters, the work of
reading, in part, consists of uncovering how caption, main text, and photographs
are linked. The work of reading is required to construct and reconstruct the orderly sense of the scientific findings and concepts presented. The text provides

'What is happening here is that if rain is falling on the land it would fall onto these mountains and it would drain either down one side or down the other side. So little creeks would form up here and they drain into a bigger creek, and into a bigger creek, and drain all the way down into this larger area'.

Figure 5.1. Example of a photograph and a lecturer's utterances that describe what the audience is to see in it; simultaneously, the description teaches the audience how to look to see what is to be seen. (Photograph © Wolff-Michael Roth, used with permission.)

particular constraints in terms of how it orders readers' ways of looking at photographs; the text therefore constitutes a particular pedagogic arrangement.

The situation is different, however, when we consider photographs in the presence of a verbal instead of a written text. Consider for instance the photograph in figure 5.1 and the text produced by the person talking about it. If this photograph and the associated text were part of a textbook, readers would certainly experience some difficulties to interpret this figure. For instance, which mountains are 'these mountains'? Where, in the photograph, is the 'other side' referred to in the text? What exactly does the utterance 'up here' refer to in the photograph? From the use of terms such as 'here' and 'this', which are inherently context sensitive, one can infer that the text associated with the photograph in figure 5.1 is not part of a textbook. If one considers the text and the photograph in figure 5.1 as part of a speech situation, such as a lecture for example, one can interpret it in a different way. In this example, the audience has to rely on peripheral clues for interpretation. It has to take into account not only the established text but also the clues provided by the context to make sense. However, even realizing that the text accompanying the photograph in figure 5.1 is not a written but a spoken text, the problem of identifying the, by the lecturer intended objects in the photograph is still present. Further resources are necessary to enable audiences to make sense of this photograph and text. In a speech situation these resources consist, among others, in the gestures and body orientations that speakers use during their talk. It is, therefore, as if the two classes of organizational materials, text and photograph, are linked through and coordinated by the gestures and body orientations. This chapter is centrally concerned with speakers' gestures and body orientations as resources produced as meaning-making (semiotic) resources for the audiences.

Talk, gestures and photographs

In a speech situation, such as a lecture, speakers and listeners make available to each other resources (body movements, gestures) that allow coordination of speech in particular, and of the entire interactions more generally. There is evidence that listeners actively interpret even highly idiosyncratic gestures and use them as resources to make sense; in fact, when speech and gesture express different concepts, gestures are usually more reliable and more conceptually advanced.[1] Given the importance of gestures in teaching and learning settings, particularly mathematics and science environments, it is surprising that the role of gestures in scientific and mathematical discourse remains largely unexplored in educational research.

Gestures have been classified into different types including beats (or batons) and gestures of deictic (pointing), iconic, and metaphorical nature. Beats are gestures that are void of propositional or topical content, and yet lend a temporal or emphatic structure to communication. Beats function as interactive gestures, which serve to regulate the coordination of speaking turns, to seek or request a response, or to acknowledge understanding. Deictic gestures are used in concrete or abstract pointing. They 'point' out some aspect of the content, making it salient figure against everything else, which becomes rather diffuse ground. The accompanying utterance about this aspect is therefore *grounded* by means of a relation to the referent that is made salient by the deictic gesture. These gestures, thus, are context dependent. Deictic gestures, coupled with deictic utterances, play an important role during interaction because they establish a distinction between figure (topic) and ground. In addition to pointing out features in the environment and indicating directions, they are also used to establish and maintain abstract spaces that become taken as shared so that speakers can make subsequent use of them without employing words.

Gestures are called iconic when their shape is isomorphic with the content they convey, that is, iconic gestures are hand/arm movements that bear a perceptual relation with concrete entities and events. This perceptual similarity constitutes their communicative strength because of a nearly transparent relationship to the idea they convey, particularly within a narrative event in which they depict co-present concrete objects and events. The hands next to the temples with fingers configured such as to suggest horns while talking about a charging bull constitute an iconic gesture. Metaphorical gestures are like iconic gestures but provide a visual expression of abstract rather than concrete objects. A mathematician, whose left palm approaches the steady right palm in the context of talk about mathematical limits (e.g., in calculus), produces a metaphorical gesture.

In communicative encounters, speakers and listeners make available to each other many resources that provide contexts for constraining the meaning of utterances. Various kinds of movements with different parts of the body provide cues on how to understand just what is being said by limiting the range of possible figures to be isolated in a photograph. The body is so important for making sense in speech situations that there is a greater likelihood of communicative breakdown and need for conversational repair if visual access is barred. During lectures that include inscriptions, gestures are important resources for the presenter to organize the alignment of talk and the visual representations. Gestures are important for sense-making processes, for when there are misalignments between features of the inscription, gesture, and speech, comprehension on the part of the audience probably is made more difficult.

Photographs, gestures and body orientation

Communication can be analyzed in terms of the dialectic (mutually presupposing) relationship between imagery and language. In the present situation, there are two modes of imagery—photographs and gestures—that constitute the counterpart to speech; together, imagery and language form a unit. Any unit that embodies simultaneously different but inseparable elements is a dialectic unit. This means that there are inherent tensions between the text (written or verbal) and the imagery embodied in photographs and gestures. These two means of communication can be contrasted in terms of the difference between the typological nature of language (words and categories) and the topological nature of images and gestures (continuity, shape, and surface).[2] However, there is more to the imagery-language dialectic, for the speech and gesture are produced in the course of talk but photographs constitute something like a stable ground. Together, therefore, talk and gesture are also in a dialectical relation with the photograph—collectively produced, they inform listeners about what might be found in the photograph. At the same time, the photograph may be treated as evidence for the existence of the phenomenon elaborated by the lecturer. The gestures co-occur with the text and are directed toward the photograph, providing anchors that integrate text, photograph, and gestures into a total performance.

The types and shape of gestures produced by a speaker also depends on body orientation. Body orientations, in fact, constitute interpretive frames for gestures limiting their interpretive flexibility and thereby enhancing and even multiplying the meaning of photographs. We therefore analyze all events in terms of gestures and body orientations.

Photographs generally lend themselves to a wide range of different interpretations. When making reference to a photograph, a lecturer can use different strategies to constrain the emergence of diverse interpretations that, though legitimate, are not suitable for the topic at issue. In a textbook or article, these strategies include the written text, the indexical reference to the photograph, the caption and, in exceptional cases, the use of arrows, highlighting, and so on drawn onto the photograph itself. These strategies limit but do not completely avoid the emergence of different interpretations of the photograph. When a photograph is used in a lecture situation, however, the strategies used by the lecturer to constrain the occurrence of misinterpretations are different. Although the text still plays an important role in directing the audience towards the 'right details' in the photograph, the ambiguity of the text implies the use of more explicit associations between text and photograph. In this context, the gestures and body orientations function not only as the reference to the photograph, but also as the means by which text and photograph are explicitly associated with each other. Through gestures, lecturers narrow the range of ways for looking at photographs. Thus, although most journalists and television audiences saw the Los Angeles police officers beat up Rodney King, one can show how the police expert called by the defense taught the judge and jury to view the video recordings as evidence for the contention that Rodney King was not 'beat up' but 'intelligently kept under control'.[3] In the previous chapter, we also showed that narratives and framing allow textbook authors to constrain how readers look and what they see in pictures. Our research shows that specific instances of the talk are directly associated with a correspondent gesture and body orientation, which, in turn, guides the viewer through the photograph. That is, the lecturers guide their audiences through photographs not only by means of text but also by their deployment of gestures and body orientations.

What utterances lack in sophistication or specificity is provided by the gestures and body orientations directly associated with them, that is, the text which by itself could generate many doubts because of its ambiguous nature, is enhanced by the gestures, which complement the text. Both text and gestures together form a structure that becomes the lens through which the photograph is viewed. Although different interpretations may still arise, a photograph in a speech situation can have its interpretive horizon narrowed such that the opportunities for new and diverse interpretations are by and large reduced.

The body orientation is part of the periphery of the gestures that contextualizes the entity to be made salient in the photograph. Such changes in periphery can be perceived as a clue that allows the audience to move its attention away from the currently projected photograph and to focus on the relation between words and gestures alone. When a speaker stands between the audience and the screen onto which photographs are projected within reach of the speaker, we distinguish two orientations. In the first, the speaker's head and frequently shoulders and upper body are oriented toward the photograph. This orientation signals that gestures and talk pertain to something visible in the photograph. In the second, the speaker is clearly oriented toward the audience. This body orientation generally signals that the current topic is about something not directly available in the photograph.

When a photograph is used in a lecture situation where photograph and lecturer are visible to the audience, the body orientation therefore becomes a resource on which the audience can rely when interpreting the photograph in the context provided by the spoken text. The lecturer's body orientation helps to distinguish different types of gestures and provides resources for connecting verbal text and photograph, insofar as it provides cues to the audience about where to focus their attention, either on the photograph or on the lecturer. Each position or orientation the lecturer assumes represents a different phase of the speech, and the audience is provided with a resource for grasping that just by looking at the position of the speaker. When lecturers turn towards a photograph, their body orientations account for the same as if they said, 'now look at the photograph'; however, when they turn sideways or fully to the audience, turning their backs to the photograph, this is the same as saying, 'now pay attention to me and my gestures but not the photograph'. Lecturers do not have to say this in words, but the attitude is expressed by changing body orientation.

Functions of gesture and body positions

The gestures appear as the most important semiotic resource that the audience could rely on to interpret the photographs in the course of a lecture. Our classification of the gestures therefore helps us to investigate the pedagogical functions a photograph can achieve when it is used in a lecture situation as opposed to a textbook or journal situation. Describing each category of gestures allows us to analyze the interactions between photograph, text, and gestures that are associated with the work of interpreting and using photographs during lectures. We emphasize the function of each gesture in the working of interpreting the photographs and connecting them to the topic of the discourse, that is, the subject matter to be taught. To limit the amount of text required to explain the concepts of the different lectures analyzed, we take all our examples from one fifteen-minute lecture of the biologist-environmentalist who repeatedly assisted two seventh-grade teachers in the process of implementing a four-month environmental unit.

In the particular lecture chosen, the biologist-environmentalist attempted to teach the concept of 'watershed' drawing on photographs, aerial photographs, and maps mounted as slides and projected against a screen covering the chalk-board. Throughout the presentation concerning the concept of watershed, the biologist-environmentalist stood next to and in front of the projection screen. She used many of the same photographs in other recorded lectures, including one that she gave to the scientists in a nearby federal research institute.

During our analysis, we classified the different functions that gestures/body orientations had in relation to the photographs when used by the different lecturers in our database. Our classification includes eight different functions of gestures/body orientations produced as semiotic resources for making sense of photographs: (a) representing, (b) emphasizing, (c) highlighting, (d) pointing, (e) outlining, (f) adding, (g) extending, and (h) positioning (table 5.1). In this classification, we take into account the position of speakers when they gesture, the relation of the gesture with the photograph (e.g., if the photograph is used as a background for the gesture or not), the visual availability of the object/phenomenon in the photograph, and the primary function of the gesture in relation to speech and photograph. We distinguish the production of gestures along the following lines: gestures are deictic, iconic, or both; they can be distinguished as specific or generic; they present a phenomenon that is or is not available in the photograph; and body orientation (table 5.1). There were a total of ninety-two gestures during the fifteen-minute lecture.

Table 5.1
Functions of gesture/body positions in the presence of photographs and speech

Function	Characteristics				
	Gestured phe-nomena avail-able in photo?	Deictic/iconic	Specific/generic	Body position	N
Representing	No	Iconic	Generic	Towards the audience	30
Emphasizing	Yes	Iconic/deictic	Generic	Towards the photograph	20
Highlighting	Yes	Deictic	Generic	Towards the photograph	12
Pointing	Yes	Deictic	Specific	Towards the photograph	11
Outlining	Yes	Deictic/iconic	Specific	Towards the photograph	8
Adding	No	Iconic	Specific	Towards the photograph	6
Extending	No	Deictic	Specific	Towards the photograph	3
Positioning	No	Iconic	Generic	Towards the photograph	2

Representing

The gestures classified in this category are those that speakers use to represent objects or phenomena not directly available in the photograph and yet are associated with some feature of it. Although utterances are related to the photograph, the gestures are about something not directly visible in the photograph such as a gesture enacting the downward slope of a road seen on the projected aerial photograph (figure 5.2). These gestures are iconic, resembling the shape or the movement of something real, familiar to the audience. They are also generic because they refer to slope in general rather than portraying something specific in

'And you go sort of through the Tsartlip Band Reserve, and you start to head
down the hill a little bit'.

*Figure 5.2. Example of representing category of gestures. The photograph does not con-
tain the feature represented in the gesture. The speaker turns the head toward the audience, but remains sideways so that the downward motion of hand/arms remains visible.
The gesture is iconic and generic.*

question. When using such gestures, the speaker is always (100 percent) oriented such that the speaker's regard falls somewhere between paralleling the
projection screen to facing the audience (making 25 percent of all possible body
orientations). This type of gesture/body orientation is the most frequent in the
watershed lesson analyzed. In the entire presentation, 30 representing gestures
have been recorded.

Figure 5.2 presents an example from the category of *representing*. In this
sequence, the speaker represents the downward slope in a road previously
pointed out by another type of gesture. Her body is turned halfway to the audience, shifting position as she starts talking about a sloping hill that is not represented in the aerial photograph. The speaker looks at the audience as she utters
and gestures, providing a noticeable frame directing attention away from the
photograph and to the gestures themselves.

These gestures, therefore, not only provided resources to focus attention,
shifting from photograph and speaker to speaker alone at the moment that the
audience is expected to do so, but they also help to understand the topic of the
discourse, insofar as they represent something not visually available or available
in another way. That is, in the present situation, the audience can see in the photograph a bird's eye view of road winding its way from the village along the
coast and toward the creek that defined the watershed. The speaker's gestures

'So this is basically a drainage area that is collecting the water that falls on the lands'.

Figure 5.3. Example of representing *category of gestures. The speaker completely turns toward the audience and provided a visible image of the basin that constitutes a watershed that drains into a creek; the phenomenon is not directly available in the aerial photograph. The gesture is iconic, representing the heights of land, and generic, representing types rather than specific heights.*

provides a means for students to connect to their experience of driving along this road; that is, the gestures constitute an iconic, concretely embodied representation of driving along the road that was perceptually available to the audience only by means of the aerial photograph. The gestures open up a third dimension in the photograph and connect it to the discourse and the real world the photograph partially represents.

The topic of the lesson—from which the examples of the different categories of gestures are taken—is the concept of 'watershed'. In this context, the lecturer's representational gestures help the students to define a watershed. The road follows a downhill slope towards the creek that gathers all water in the area; in fact, the road crosses the creek only two hundred meters from where the latter sheds into the inlet. The gestures are used to provide a visual image of the downhill slope, which the photograph by its nature cannot depict. Similarly, when the lecturer bends her arms, forming a circle parallel to the floor, at the same time referring to a watershed, the audience is able to associate visual and verbal resources to define a watershed as a delimited area (figure 5.3). Here, the two-armed gesture (e.g., figure 5.3) produced in the context of an orientation towards the audience signifies the 'heights of land that define and delimit the

'Graham is coming up this way and Hagan is coming down this way'.

Figure 5.4. Example of an emphasizing *gesture: The referent entity is available in the photograph, the gesture is generic and iconic/deictic, and the body orientation toward the photograph.*

watershed'. These heights of land, in contrast to the creek, are not visible in the projected image, a fact clearly signaled in the corresponding body orientation. This second instance, therefore, despite its perceptual similarity to the gesture in figure 5.4, is a representing gesture. In another situation, she uses the same gesture but then moves both arms downward until they meet and then meanders to suggest the creek flowing into the ocean inlet.

Emphasizing

In this category, we count iconic gestures that emphasize an entity directly available in the photograph, by generically following the shape, movement, or direction of the object (phenomenon) referred to in speech; the gesture therefore also has deictic function. When speakers gesture in such cases, they generally are positioned towards the photograph (shoulders somewhere between parallel to the audience's line of sight and parallel to projected image) and the gestured phenomenon is available in the photograph. Here, too, the speakers' orientations functions as a frame orienting the audience to look at the relation between gesture and corresponding features in the photograph. In the lecture featured here, there were twenty emphasizing gestures produced during the presentation on the watershed concept.

In figure 5.4, for example, the gestures emphasize the confluence of two creeks (Graham, Hagan), and how they come together at some point. The background for this gesture is an aerial photograph, and, although the speaker is not specifically tracing the creeks in the photograph, she approximately represents

'And so up here, any idea [what this is]?'

Figure 5.5. Example of highlighting. *The gesture is iconic, pointing to something, but generic because the outline of the thing is not specifically identified. The orientation is toward the photograph.*

the direction of those creeks in the way they are perceptually available in the photograph. The gesture is iconic, for there were two creeks that come together at about the area she gestures; the gesture is generic in that neither right nor left hand arms parallel the creeks; and the gesture is deictic, for it allows the students to look in a particular direction for finding the referent of the gesture.

Central to understanding the concept of a watershed is the idea that all water falling within its confines is carried away through a river system that sheds at one location into the ocean. A watershed is a drainage basin common to a particular area, defined by 'the heights of land'. Figure 5.4 clearly shows that the gestures do work in emphasizing the common direction that two different creeks in the system are taking. After pointing out the two creeks sequentially, the double-arm gesture simultaneously represents the two creeks and their confluence, which emphasizes the aspect of a watershed as a drainage basin. When she turns away from the photograph and rotated the double-armed gesture into the horizontal direction, the same gesture represent the 'heights of the land' from where the water flows and represents the two creeks that meander (meandering motion of hands joined at palms) downwards to the ocean. This downward flow of water in a watershed is not visually available in the photograph, thus the rotation away from it. They become perceptually available, however, through the representational gestures. The gestures here add the ideas of downward flow and motion, and therefore link the static photograph to a dynamic phenomenon.

Highlighting

Highlighting gestures are deictic but have a generic shape; they usually are circular or elliptical in shape without having clearly determined boundaries. These gestures are used to focus attention to the approximate area where something was to be found; the orientation is therefore toward the photograph. For example, while introducing students to an aerial photograph of the watershed that also included their school and village, the present speaker uses these gestures to direct attention to different but not well-defined areas that she wants students to identify, such as the mountain that dominates the valley. Because there are no determined boundaries, it is harder to identify the object completely; the circular gesture (e.g., figure 5.5) simply directs the viewers' attention to some area in the photograph, but there are no details available that assist the audience in identifying the specific feature to be attended to. It is therefore not surprising that it takes several increasingly concrete prompts before the students provide the name of the entity (mountain, quarry) that the speaker wants them to identify. There are twelve highlighting gestures to be found during the mini-lesson on the watershed concept (table 5.1).

We consider these gestures as a special type of deictic gestures, insofar as they point out something in the photograph, but in a very general way, without clearly identifying boundaries as *outlining* gestures (see below) do. In this sense, they are deictic gestures that highlight something in generic fashion instead of specifically pointing to something in the photograph.

Photographs are inherently full of details many of which are not relevant to the main features; this increases the difficulty in selecting just what is to be perceived in the photograph as we see in the present case when students named features other than the mountain that the speaker wants to make salient. There are almost endless possibilities of interpretation of the same photograph. The text narrows these possibilities and gives the audience guidance towards the 'right details' in the photograph, that is, those details required for understanding the watershed concept. In a textbook, caption and body present the relevant *con*text. In a lecture, the text is provided by verbal utterances; gestures and body orientation of the lecturer constitute additional semiotic resources available to the audience that in fact will help them to distinguish background and foreground in the photograph and choose the right details as the relevant ones in any given context. When highlighting the mountain in the photograph, the speaker directs the attention of the audience to an essential object in the photograph. A watershed is delimited by the heights of the land, and in this particular watershed, the mountain in the photograph is part of the boundary that defines the watershed within

'And your school would be right here'.

Figure 5.6. Example of pointing. *The gesture is specific and deictic and associated with a body orientation towards the photograph.*

which the students' village and school lies. Therefore, identifying this object in the photograph, even if only in generic fashion, is an important part of the lesson and highlighting gestures assisted, though could not entirely disambiguate, the nature of the watershed boundary.

Pointing

Pointing gestures belong to the class of deictic gestures. During the lectures analyzed, speakers pointed to specific objects in the photograph, or, in some instances, to the entire photograph. Pointing is very specific, towards an object in the photograph that is clearly defined (at least to the speaker). It is also frequently accompanied by deictic terms such as 'this', 'that', or 'here'. The objects in this situation are always visually available, that is, the action of pointing requires the availability of the object in order for the audience to understand. This characteristic of the object—its visual availability—is the most important one to distinguish between what is considered a pointing and what is considered an extending, according to our classification. In the lesson about watersheds, there were eleven pointing gestures.

In figure 5.6, for example, when the speaker points to the photograph and said, 'right here', she did more than just pointing. That is, the students have available not only an index for finding their school in the photograph, but also an indication of the boundaries of this object, its extension—it is small in comparison to the entire photograph or extended objects such as the mountain in figure 5.5. When the extensions of the entity are small, students can easily identify what is being pointed out. The gesture brings the exact location of the school in the aerial photograph to the foreground, leaving everything else in the background.

These gestures are distinguished from the previous category (highlighting) because of their specificity; that is, pointing implies a very specific object or phenomenon in the photograph, while highlighting is a more general gesture, pointing out something not entirely delimited in the photograph. In this situation the possibilities of different interpretations of the photograph are narrowed such that only one interpretation is possible, namely that intended by the speaker. As long as the audience is able to see what the lecturer is pointing to in the photograph, the connection of text and photograph is immediate. However, when the entity to be referred to extends in space, a simple pointing gesture is insufficient as it might refer both to the entity and one of its parts (which a novice could in fact interpret to be the entity). In this situation, outlining gestures provide semiotic resources to the students.

Outlining

Outlining gestures are very specific deictic gestures often used to follow the shape of some entity in the photograph; because over time the gesture traces out a shape (nearly) identical to the object, it is also iconic. In outlining gestures, the speaker always makes use of the photograph, which the audience, assisted by the gesture, is expected to divide into figure and ground. The shape of this gesture depends on the visual availability of the object in the photograph insofar as that what is outlined is something in the photograph, visually available to the speaker and to be identified by the audience. In the talk about the watershed concept, there are eight gestures that can be found outlining an entity in a photograph.

Figure 5.7 exemplifies the outlining of a specific area in the photograph. As expected, the speaker is directed towards the photograph, carefully following the shape of the coastline, thereby defining the boundary of the inlet into which the watershed empties. Both her gestures and the referent object (Saanich Inlet) are visible from the audience's perspective. Because of the close spatial relation between the moving pointer (finger), the possibilities of mistaking the coastline and with it 'Saanich Inlet' for something else are greatly narrowed and the identification of the relevant details in the photograph is immediate. In following the coastline, and thereby in outlining the inlet in the aerial photograph, the lecturer provides a resource for students to guide their attempts of connecting the photograph with their lived and experienced world, their neighborhood. In doing so, the speaker provides a concrete part of the complete system that defines a watershed. The gestures constitute important resources in isolating the concrete case

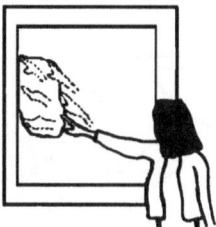

'So this would be Saanich Inlet over here'.

Figure 5.7. In this example of outlining, *the speaker used her index finger to follow the coastline that defined the ocean inlet: the gesture is specific, deictic-iconic and the body orientation is toward the photograph.*

of a whole watershed or its parts from the interpretively under-specified photograph.

Adding

Adding gestures are also used to outline entities, but, in this case, the object or phenomenon is not visually available in the photograph but *could have been* there. The gestures provide another layer of specific perceptual objects that are created in iconic form in front of the photograph but are to be understood as an addition. Gestures that add something to a photograph occur six times in the watershed lesson.

Consider for instance figure 5.8. In this example, the speaker models a phenomenon, 'an oil spill in the heights of the land', as it might unfold in the area represented in the photograph. As she talks about a potential oil spill, she traces what would be the results of this hypothetical phenomenon in the photograph, that is, the oil flowing down into the creeks and subsequently into the lake. Although neither the creek nor the phenomenon of an oil spill can be seen in the photograph, the speaker 'draws' another, virtual image of oil flowing down into the creeks and into the lake. The situation is hypothetical—but the gestures render this event concrete. The speaker adds something, literally layering it onto the photograph in a way that only her gestures can make it perceptually available to the audience. This is a completely new semiotic resource for understanding pho-

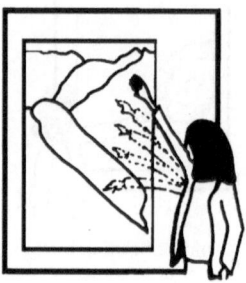

'Say something happened up in the heights of the land, the headwater of that
area, like an oil spill. You would ultimately be able to trace the impact of an oil
spill up in the top, all the way through the creeks, and its impact right down in
to this lake here'.

*Figure 5.8. Example of gestures that add something to the photograph. The gesture is
specific/generic and iconic/deictic, associated with a body orientation toward the inscription.*

tographs not achievable in a textbook. The gestures can be understood as a form
of concrete, public, and witnessable thinking.

In this example, the presenter emphasizes a particular feature of a watershed,
that is, the idea of a watershed being a drainage area that is delimited by the
heights of the land and providing the topology defining water flow. By adding a
new event using the photograph as a background, a direct association is established between the photograph and the idea of a watershed, as well as the consequences of pollution to a watershed; this pollution would not just remain where
it occurred but, by moving in the way the gestures showed, would have an effect
on the entire system. The lecturer here does not only associate photograph and
speech for defining a watershed, but also amplifies the photograph as if it were
showing an event.

Extending

Extending gestures are specific deictic gestures used to add something to the
photograph that does not fall within its boundaries. That is, the entity referred to
is located outside the limits of the photograph, so that, if the photograph had
been taken from a greater distance, the entity would have been included; conse-

'Sidney is further up this way'.

Figure 5.9. Example of gestures that extends the photograph, pointing at something specific somewhere off the photograph and therefore not directly available but would have been visible if the photograph had been larger; consequently, the body orientation is toward the photograph. The gesture is specific and deictic.

quently, the body orientation is toward the photograph. In this sense, the chalkboard, the wall, and anything that surrounds the photograph becomes an extension of it, insofar as the speaker pointed out something in these areas, outside the photograph. Because the gesture adds something beyond the boundaries of the photograph, the speaker orientation is, consistently, toward the representation (figure 5.9). In the watershed presentation, these gestures are observed three times.

As with the adding gestures, the audience is invited to imagine something not visually available in the photograph. The gestures connect the photograph, as a representation of something real, to the actual real world, extending the boundaries of the photograph to include other aspects. Talk and gestures are again a means by which the lecturer can transform the photograph as to show to the audience something that is actually not visually available just by looking at it. It is almost as if the photograph is a different photograph; the lecturer points to objects that are outside of the photograph just beyond of its boundaries and yet as if represented on the (extended) screen. The audience is invited to follow her through her gestures and speech to envision the aerial photograph as covering a little more area. The relevant function of this gesture to the effort of defining a watershed is the fact that the lecturer can introduce additional elements that assist the audience in understanding the concept or the location of the photograph with respect to the larger setting known to the audience.

'So, we are standing on the southern boundary here'.

Figure 5.10. Example of positioning, *a specific/generic gesture of iconic type that, together with the body position, virtually extended the photograph into three-dimensional space.*

Positioning

This type of gesture, specific/generic and iconic is strongly related to body orientation; it constitutes another type of extension, but this time into three-dimensional space. Speakers position themselves against the photograph as if they were actually taking the shot at that exact moment, standing in the landscape depicted. Speakers therefore allow the audience, through their body orientation, to understand the photograph as if it is extended into the lived space to produce a three-dimensional image. There are two examples of positioning in the lesson analyzed, both in the context of landscape photographs.

Figure 5.10 provides an example of positioning. The resulting representation goes one step further than any photograph—through its association and placement relative to the speaker it makes the photograph an extension of the watershed into the classroom allowing the speaker's actual movement to become movements in the watershed. The speaker in fact takes up a position in a hybrid world that instantiates the watershed in the classroom. Positioning works as an explanation, an introduction to the photograph itself and the topic related to it. The lecturer introduces the photograph to the audience as a depiction of the real world, making a direct connection between photograph and real world in showing how and from what angle the photograph has been taken. By doing this, she also makes explicit the role of the photographer in the production of the photograph, exposing the human interference in this process that could otherwise be regarded as an essentially objective representation of the world.

The introduction of the photograph to the audience provides resources for looking at the photograph in a particular way, and it is very important for the work of interpretation of the photograph that will eventually follow. In this particular case (figure 5.10) the lecturer identifies the boundaries of another watershed, where the heights of the land that define the area are more difficult to be identified in the photograph. Therefore, she positions herself such as to simulate where and how the photograph was taken so the audience would be able to recognize the photograph as a depiction of an area they are familiar with. She subsequently shows how the creek flows, first in a direction perpendicular to and approaches the areas seen in the photograph, then turning and flowing parallel to the screen toward the ocean on the left (from the audience). It is immediately evident that there are no examples of positioning when speakers talks about similar photographs but does not stand near the projected image (e.g., the biology professor using transparencies or the environmentalist using laser pointer in a darkened lecture hall).

Graphicacy: constraining, amplifying, and multiplying meaning

School science relies heavily on inscriptions such as photographs as pedagogical elements in teaching facts and concepts; they are the most frequent inscriptions in secondary-level biology textbooks. Whereas we show in chapter 4 the limited number of semiotic resources available to high school students when facing photographs in Brazilian textbooks, the present chapter shows how hand gestures and body orientations provide additional semiotic resources that teachers and lecturers can make available to assist students in making sense of photographs and the new concepts that they pictorially render. These gestures and body orientations play an important role in understanding just what students are to attend to in a photograph, whose meanings are inherently under-specified; they are therefore an integral part of graphicacy in lectures. Understanding the relationships between talk, gestures, and photographs in science lectures therefore should be an important consideration in developing pedagogy related to critical graphicacy in general and the use of photographs in particular.

For the situation where teachers or university lecturers stand next to the photographs projected onto a screen in a room where they and the gestures are clearly visible, we identified eight categories of gestures. These are distinguished by the function they have in relation to the photograph and exemplify these categories in the context of part of a talk concerning the concept of 'watershed' presents by a biologist-environmentalist to an audience of seventh-grade students and their teachers. These categories, which are important component of graphicacy in lectures, include representing, emphasizing, highlighting, pointing, outlining, adding, extending, and positioning. All eight types constitute important resources that characterize knowledgeable graphicacy in addition to verbal literacy for understanding photographs, their contents, and their relevance to the theoretical concepts to be learned. They do multiple duties by limiting, amplifying, and multiplying meaning.

These eight types of gesture and body orientation combinations constitute a range of semiotic resources not normally made available in textbooks. It is possible that written texts could provide suitable textual information to do what gesture and body orientation do. But identifying a specific object in an inscription by means of a written text would require a very descriptive and detailed text to generate a similar explicitness of the relation between text and inscription. The written text would become much more complex, and, yet, would not be able to guarantee that the relation between text and inscription will be appropriately established by the reader. Previous research shows that there are many features that make it difficult to understand photographs in the way they are deployed in textbooks. With present technology—photocopying, scanning—teachers could actually project textbook images and, deploying relevant gestures and body positions, could assist students in picking out relevant detail to assist them in meaning-making processes. This would make the highly text-oriented way of teaching science to include other modes of representing facts and concepts, including inscriptions and gestures. It would be interesting to study whether students' levels of graphicacy increase when they are exposed to whole-class presentations and readings of inscriptions where the teacher uses gestures and body orientation explicitly as additional resources for making salient or adding features.

As additional semiotic resources, gestures and body positions take advantage of the fact that they are of a different kind than text. If a textbook were to use the words 'here' and 'there', these would be far more difficult to understand than if a speaker who used gestures. These indexical words imply a specificity that cannot be achieved by the text, at least not in the same way in which it is implemented by gestures in the speech situation. Not only the inscription itself would have to have additional signs that would delimit the object of interest but also the text would have to be more detailed and descriptive to identify this object in the photograph. Gestures and body orientation are of a different kind than written and pictorial resources and, by their very nature, can be layered onto the inscription without nevertheless encumbering the auditory information channel. That is, because gesture and body orientation are of a different kind they can be apprehended without requiring additional attention resources.

In a speech situation, the periphery of the text—i.e., its *con*text—is enlarged by gestures and body orientations, allowing the use of fewer words in a less structured way, and still being able to transmit the same message; the different semiotic resources multiply meaning because, in some sense, they include a certain level of redundancy in and synergy of the different modalities. Classrooms where the gestural modality become available to teachers (and students) in addition to photographs and texts therefore implement some of the recommendations educational psychologists have made with respect to the need to use multiple inscriptions of the same concept; gestures and body positions are a means that make classroom teaching inherently adaptive to the graphicacy needs of students.

Much of the simplicity of the verbal text is due to the presence of the gestures that accompany the discourse. The gestures also function as part of the text (discourse), as a resource that simultaneously simplifies and amplifies the text, and consequently, enhances the explicitness of the connection between text and inscription. The gestures and body orientations that accompany speech and are directly related to the inscription allow the verbal text to be simpler and yet more specific and constrained in meaning than the written text. That is why, in some instances, we expect students with difficulties understanding concepts in the book to more easily understand the same concept when the teacher (or someone else) provides them with a very similar explanation.

Role of spatial configuration

There is evidence that classroom discourse changes in fundamental ways when gesture and body orientation become available as resources to teachers and students alike because the spatial configuration allows speakers to stand next to the artifacts and inscriptions.[4] When speakers do not stand next to the artifact or inscription, the communicated content becomes more limited and the audience understands less. Thus, spatial configurations that do not allow the deployment of gestures and body positions to complementing speech decrease the number of semiotic resources and become similar to those available for understanding inscriptions in textbooks. For instance, consider the differences in two situations when we recorded an ecologist talking about watersheds first in a classroom where her gestures and body orientation are visually available to the audience and then in a dark auditorium, where the audience can only listen to her and look at the photographs projected in the wall. In these two different lecture situations, the role of the inscription and the discourse itself changes drastically. When the gestures are visually available to the audience, the text is simpler and more specifically directed towards the inscriptions; the inscription becomes the topic of the discourse, and its pedagogical function in helping to understand the scientific concept presented is enhanced. However, when the gestures are not visually available to the audience, the lecturers (in biology course, environmentalist speaking in darkened lecture hall) only make references to the photographs while they speak using pointers but the gestures are not available to make sense of what the text was about. Details pertaining to the inscription but which are not visible cannot be made available in the same way that the environmentalist has done in the examples provided here.

When lecturers use transparencies, the gestures are made toward the transparency, and, although they function to highlight some aspects of what can be seen in the inscription on the transparency, the lecturer's body orientation, as well as all the other types of gestures that do not use the photograph as a background, are absent from the audience's view. Therefore, the audience members have to attend to what is being projected on the wall without other semiotic resources being available in the case of talk next to the projected image. The fact that the room must be dark and the fact that the lecturer gestures on the transparency instead of on the projected image on the wall, further contributes to the distancing of the audience from the lecturer and separating him or her from the photograph. In this case, the gestures that do not make use of the inscription as a background and that are in other circumstances important resources to facilitate the interpretation of the inscription are not available. The same types of gestures are not used, or if used, are not visually available to the audience, limiting what could be learned from inscription had the additional semiotic resources been available as a pedagogical means. That is, the importance of an inscription as a pedagogical resource in a lecture situation is greatly increased when the audience is able to see both inscription and lecturer at the same time, and when the lecturer gestures over and about the inscription, thus exploring the inscription in its full potential. In the potential chaos arising from the proliferation of new ways of talking that students may experience in lectures, in which the same words are often used to denote different objects, gestures and body orientations are crucial resources for establishing a coherence that allows audience to appropriately connect inscription and text, being able to understand what is being talked about. Gestures, therefore, play an important role in science discourses, and should be more carefully investigated.

6 Interpretive graphicacy in practice

Figural representations, that is, inscriptions, may be worth a thousand words; on their own, however, they mean very little, as we showed in our Introduction and in the previous investigations. Thus, inscriptions give rise to innumerable, different interpretations because their meaning, that which is shared in a community, emerges from mutually presupposing relation between the author's way of seeing and the perceptions of the reader. This is just what John Berger brought out in the quote at the beginning of this book with respect to one type of inscription, photographs: 'The photographer's way of seeing is reflected in his choice of subject. . . . Yet, although every image embodies a way of seeing, our perception or appreciation of an image depends also upon our own way of seeing'.[1] That is, graphicacy has two sides to it. On the one hand, there is an author (photographer, designer) who chooses certain dimensions or features to be highlighted; these choices enter the making of the inscription and subsequently frame, without being able to limit, potential readings. On the other hand, there are readers, who also display graphicacy in and through their reading. When readers are not familiar with particular inscriptions yet graphically literate individuals, that is, individuals displaying knowledgeable graphicacy, they enact interpretive graphicacy. It is the reader's work of reading, the viewer's perception of the narrative and perceptual order of the inscription and the surrounding text, and the meaning-making resources available to the reader that allows a specific interpretation of an inscription to arise.

In chapters 4 and 5, we are concerned with articulating levels and structures of graphicacy embodied in (a) textbook authors' use of photographs and (b) lecturers' productions of semiotic (meaning-making) resources over and about photographs. In the present chapter, we focus on the other side of graphicacy, the levels of graphicacy displayed by middle and high school students when encountering photographs embedded in various amounts of accompanying text. The questions we answer here include, 'What then do high school students perceive when they look at photographic images in biology textbooks?' 'How do they use other meaning-making resources that a text makes available when they want to understand the things that the photograph is about?' More specifically, we provide answers to the following questions: (a) 'What is the role of semiotic resources co-deployed with photographs?' (b) 'Do students 'read' photographs when they are studying a text?' (c) 'Do students' interpretations of photographs change when they are provided with additional text?' (d) 'How do interpretations change when other meaning-making resources are provided?' and (e) 'Is there a difference in the way students interpret multiple versus single photographs?'

Background

To investigate photographs and their relation to captions and main text during interpretation, we interviewed students from two distinct groups: (a) students who did not have an ecology course at the high school level and (b) students who already studied ecology as part of their compulsory high school biology course. With this choice, we also wanted to determine the role prior knowledge may have on the interpretation and comprehension of photographs and associated texts. In most of Brazilian schools, primary education includes grades K through 8, and high school includes Grades 9 through 11. Basic notions of ecology are normally taught in general science courses that students attend in fifth or sixth grade. An expanded ecology curriculum is taught as part of the biology curriculum while students attend grades ten or eleven, depending on the school. We interviewed twelve students, six from each of the two groups, who attended either private or public school.

We used four photographs during the interview based on the analyses in chapter 4 of the different functions they played as meaning-making resources. The following distinctions were addressed in the selection of the photographs: (a) there were both single and multiple photographs; (b) some photographs were referred to in the text others were not (i.e., incidence and placement of a feature such as 'Fig. 30.3'); and (c) all four major categories of photographs identified in chapter 4 had to be represented. We began by selecting a variety of photographs that could address one or more of these aspects.

The 'orchid' photograph represents a single photograph and is associated with an indexical reference appropriately placed in the main text, and represents *illustrative* photographs (figure 6.1). The photograph has the potential to give rise to misinterpretations due to the enormous amount of details present in the photograph and due to the lack of information in the caption as well as in the photograph itself that could otherwise have helped the readers in the work of in-

Inquilinism

It is an inter-specific harmonic association, in which only one species is benefited; nevertheless the beneficiary does not cause prejudices to the other associated species.

The *tenant* (beneficiary species), gets shelter (protection) or even support in the body of the host species. It is the case of the interaction between orchids and bromeliads and the trees in which trunk they are installed (Figure 83.1). The orchids and bromeliads, differently than what some people believe, are not parasites, since they do not cause any kind of damage to the host plants.

They have adapted to live in the top of the trees, where they find ideal conditions of luminosity for their development. Therefore, they are called epiphytes (epi: above); this kind of inquilinism is also known as *epiphytism*.

Fig. 83.1 – Epiphyte plant.

Figure 6.1. Photograph of the orchid. Reproduction of the main text and the photograph with caption, as they originally appear in the textbook. In the original, the text was in Portuguese and the photograph in color. (Reproduced with the permission of the copyright holder.)

terpreting this photograph (see chapter 4). Furthermore, this photograph is one amongst very few photographs that presented some kind of semiotic resource directly placed on the photograph itself (in this case, the word *Orquídea* [Portuguese, 'orchid'] in the bottom right corner of the photograph).

The second photograph is a single photograph and exemplified *decorative* photographs (figure 6.2); that is, it lacks caption and indexical reference in the text. For the interview, we reproduced the entire page where this photograph ap-

CHAPTER 2

ENERGY AND MATTER IN THE BIOPHESRE

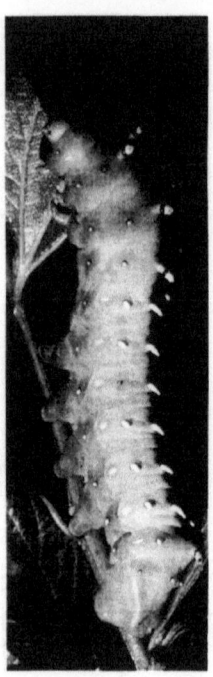

What this chapter is about

A great contribution of Ecology to the contemporary thought is to call attention to the intricate network of relations that exists between living beings and the environment. The human species, besides being part of this network, has caused a great impact on it. To know basic concepts of Ecology is indispensable to everyone who wishes to be conscious and responsible citizens.

Highlights

– The organizational levels that constitute the biosphere: ecosystems, communities and populations.
– The relations that exist between autotrophs and heterotrophs and their importance to the perpetuation of life.
– The structure of food chains and webs, and the role of producers, consumers and decomposers in the ecosystems.
– The pyramids of energy and biomass that illustrate the behavior of matter and energy in the biosphere.
– The cycles of the most important chemical elements that are essential to living beings: cycles of carbon, oxygen and nitrogen.

Figure 6.2. Photograph of the caterpillar. Reproduction of the entire page of the textbook where this photograph originally appears. In the original, the text was in Portuguese and the photograph in color. (Reproduced with the permission of the copyright holder.)

pears, leaving it to the participants to decide whether the text pertained to the 'caterpillar' photograph or not. The third item consisted of a series of three photographs; the main text did not refer to them. It represented *complementary* photographs (figure 6.3). These photographs deal with the concept of camouflage. The fourth and final photograph (figure 6.4) is a single photograph to which the main text refers in an appropriate manner; it represents *explanatory* photographs. The topic of the text is *mutualism*, and the photograph presents lichens as examples of associations representing mutualism.

Our major interest in this chapter is related to the role of photographs, captions and texts in the actual process of reading a textbook. We therefore decided

Camouflage and Mimicry

Among many kinds of adaptation, deserve highlight those that make individuals of one species become less visible and blend themselves with some things in the environment or even make themselves similar to living beings of different species. With such abilities, these individuals can hunt their preys more easily or, differently, escape from the attack of their natural enemies. These kinds of adaptation are called camouflage and mimicry.

Many insects, reptiles, amphibians and birds have green color and, thus, they make a perfect camouflage among the leaves where they hide in. Among insects, some has acquired, during the evolution, color and shape of aculeo (the false thorn of roses). These insects try to have advantage with this adaptation living among plants that have aculeos.

Fig. 577. The European partridge, during the winter, shows white plumage, blending with the snow. At the end of the winter, it starts to change its plumage, and acquires a coloration that blends with the dry vegetation where it lives. This is a good example of camouflage.

Figure 6.3. Photographs of camouflage. Reproduction of the main text and the photographs with caption, as they originally appear in the textbook. In the original, the text was in Portuguese and the photographs in color. (Reproduced with the permission of the copyright holder.)

to make use of two different strategies that allowed us to better investigate these roles. We presented photographs to the students in the following sequence: First, the interviewee received a colored copy of the photograph of the orchid. No caption or text was given to the interviewee at this point. The interviewer then asked the interviewee to talk about the photograph. Second, the interviewer provided the interviewee with the caption that originally accompanied the photograph of the orchid. The interviewee was asked to read the caption aloud. New questions were asked. Finally, the interviewee received the main text associated to this figure, and was asked to read it aloud. More questions were then asked.

Students were then presented with the caterpillar photograph. In this case, we provided students with a colored copy of the entire page of the textbook

Mutualism

Mutualism is a relation in that the species benefit themselves reciprocally, but, differently from the proto-cooperation, the co-existence is indispensable to the survival of the associated species.

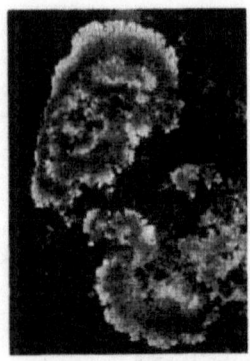

The lichens (Figure 4) constitute a mutualistic association between algae and fungi (or between cianobacteria and fungi). The fungi protect the algae, given them support, water, and mineral salts, creating conditions to the algae to do the photosynthesis; the food produced by the algae is shared with the fungi. Separated these fungi and algae could not survive.

Nowadays the expression symbiosis defines an intimate association (harmonic or non harmonic) and includes mutualism, commensalism, and parasitism.

Figure 4. Macroscopic aspect of lichens.

Figure 6.4. Photograph of lichen. Reproduction of the entire page of the textbook where this photograph originally appears. In the original, the text was in Portuguese and the photograph in color. (Reproduced with the permission of the copyright holder.)

where this photograph originally appeared. The students began reading the text or commenting on the photograph. When the students had completed, the interviewer asked additional questions from the interview protocol. Camouflage photographs were presented next. All three photographs were presented as a set following the same strategy as with the orchid photograph. Finally, the lichens photograph was presented. In this case, students were provided with the reproduction of the entire page of the textbook from where we had culled it. Again, students could either read the texts accompanying this photograph or comment on the photograph.

By presenting the photograph of the orchid and the multiple photographs about camouflage according to the first strategy we described, we expected to be able to follow the reasoning that takes place when students are faced with different kinds of information, different kinds of texts that complement one another: in this case, photograph, caption, and main text. While some students read the entire main text without interruption, other students commented while reading the texts. The intention of our second strategy (photographs of caterpillar and lichens) was to present the photographs as they might appear in the textbook, allowing us to study the meaning-making resources that students 'naturally' identified and how they used them in their interpretations.

Even though we developed an interview guide and were careful to avoid questions that could direct the interviewees towards particular features in the photograph, the nature of the interview and any word that the interviewer says has to be considered as a potential meaning-making resource to the participants.[2] If students used some such aspect in their reasoning, we would expect that it directly or indirectly shaped the answers given. We address this particular issue and provide some examples of it throughout this chapter.

Surplus of meaning from additional textual resources

'The picture helps to understand the text as much as the text helps to understand the picture'. (Faith, tenth grade)

Through our analysis, the mutually presupposing nature of the relationship between photographs and its associated texts is evident. The various semiotic resources available to the readers when reading textbooks where photographs are used were identified and carefully assessed from the students' responses during the interviews. The most important aspects we derive from our analysis are (a) the changes in how students perceived photographs when additional text was provided, (b) the work of reading when texts and photographs were provided simultaneously, and (c) the influence of the interview context and prior knowledge and experience on students' graphicacy. In this and the subsequent two sections, we articulate and provide data for each of these aspects. In this section, we show how the sequential addition of captions and main texts, respectively, mediates the levels of interpretive graphicacy that students display in the recorded interviews.

Photographs

The students' immediate reactions to the photographs were in accordance with what they had been asked. The students were concerned about providing an appropriate answer to the request 'tell me what you are seeing in this photograph' by means of finding focal points in the photographs. Their responses were constrained by the interviewer's way of phrasing the activity, insofar as asking, 'what are you seeing in this photograph?' implies a different context than if the question had been, for example, 'What is this?'

Photographs have an enormous amount of detail, compared, for example, to diagrams or graphs that make use of empty, white surfaces. This abundance of detail lent itself to a proliferation of interpretations not only between, but also within students. Thus, during their observation of the photographs, the students pointed out many other aspects of the photographs. In the photograph of the orchid, the majority of the students (ten students) pointed out the presence of many trees in the picture. This was also the detail with most variation in the term used by the students to refer to it, for example, vegetation, forest, garden, park, and so on. The next most cited object was the central tree in the photograph (nine students). Five students identified the lichens in the trunk of the central tree; and four students identified the yellow flowers on the right side of the photograph. Two students pointed out the grass on the ground in the photograph, and only one student identified the sunlight in the top of the photograph. Six students mentioned a plant, which we infer as being the orchid in the picture (one student referred to 'parasite plant', and another student used the term 'fern'). Six students actually mentioned the presence of an orchid in the photograph. When asked about what could be the topic of this photograph, three students suggested that the photograph represented 'many trees'. Four students mentioned the central tree as the topic of this photograph, whereas five students identified the topic as being the orchid. That is, rather than seeing the photograph as a concrete example of mutualism, which it represented in the textbook, the predominant perceptual feature that students highlighted were the trees, not the relationship between the central tree and the orchid.

In a similar way, students identified many different entities in the photographs that represented camouflage: Five students identified the environment around the animals in the photographs, and four students noticed the difference in the plumage of the birds in the three photographs (different feathers, different colors). Three students referred to a difference in seasons in the photographs (also referred to as difference in temperature or climates). Almost all students attempted to describe each photograph separately, emphasizing differences between the birds, the environments or the seasons represented in the three photographs. That is, rather than seeing in the series of photographs a concrete example of camouflage, students identified a variety of other features.

To actually see camouflage in this series of photograph, the viewer has to perceive what is invariant across the three images in the face of evident variations. That is, the viewer has to see the three evidently different birds as the same entity, which adapts to seasonal changes in more or less the same environment. That is, although neither birds nor environments are perceptually the same, they need to be seen as representing the same. Seeing sameness in the face of difference cannot be taken for granted—in fact, the viewer has to be instructed to be able to view photographs in this way. Seen in this light, it is not surprising that the differences in environment and in the bird itself were the most salient details across the three photographs: This is an aspect of the *productive* graphicacy that enters the photographer's work. Other details, such as the fact that a white bird is in a white environment, or a black-and-white bird is in a similar environment—specifically what constitutes the phenomenon of camouflage—did not catch the attention of the majority of the readers. The students, instead, perceived the changes in the photographs, the differences in the environment and the differences in the plumage of the bird; half of the students even said the topic of these photographs was the environment. Two students named the birds to be the topic of the photographs, whereas two students said it was the birds in relation to their environments. Only two students said the topic of the photographs was camouflage, and they were the only ones to point out the fact that the birds in the first and in the third photographs were similar to the environment, and that this would make it more difficult for predators to find the birds.

Thus, in considering the first, leftmost photograph of the series (figure 6.3), the students identified a white bird and referred to the environment as snow, ice, or simply as 'cold'. The third, rightmost photograph of the series was said to represent a brown bird in a field or mountain (students also mentioned 'grass', 'rocks', and 'hot climate'). The second, middle photograph generated a greater variety of answers. About the bird in this photograph, one student said it was black-and-white, another said it was more-or less white, and yet another student said the bird was tiger-like. But the majority of the students did not refer to the bird, describing only the environment in this photograph. The environment was identified as a river or running water, and also as intermediate temperature; only one student said it was melted ice. One student said about this second photograph that it was changing with the heat, although it is not clear if the student was referring to the bird or to the environment, or even to both. Three students also said that they could not understand this second photograph in the series.

It is, of course, interesting that a textbook in Brazil, where hardly any student will have encountered snow, would use photographs were snow constitutes an important component of the concepts 'camouflage' and 'adaptation'. Given that we already articulated the context as an important ingredient of interpretive graphicacy (see chapter 3), it is not surprising that the levels of graphicacy displayed by these Brazilian students would be mediated by the lack of experience with this phenomenon. The difficulty the students had identifying the environment in this photograph as the transition between winter and summer may be due to the fact that in Brazil the winter is very mild, and none of the students in-terviewed had ever seen snow or frozen lakes and rivers other than on television or books and magazines. Therefore, students' life experiences influenced their interpretation of the photographs; thus Brazilian students did not easily identify aspects that a Canadian student might easily identify in this photograph.

Characteristically, a series of photographs invites the readers to pay atten-tion to the differences between each photograph, by means of comparison. In a series of photographs the isolated figure takes on a meaning only in an external system of comparison; the internal comparison of the object with its background is scarcely informative. Through the work of comparing, the reader distinguishes details that differ across and those that are invariant in the photographs. By fo-cusing their attention on variations between the three photographs, the students actually missed the invariant aspects (or, at least, in two of these photographs) required to see the concept of camouflage.

In the photograph of the orchid, the students could only draw on internal comparisons. Therefore, the majority of the students opted for the criteria 'fo-cus' as relevant means to identify the important objects in the photograph, in this case distinguishing 'background' and 'foreground' in the photograph. Whether the relevant details were in the background or not is completely up to the reader in the present situation. Although common sense suggests that a photograph as a depiction of something will primarily focus on the very object that it is trying to depict, the same photograph can be used for very different and even opposite purposes, and the attention can be drawn to any detail, other than the one that seems to be the focal point in the photograph.[3] For photograph similar to the one in figure 6.5 was taken at the same spot by environmentalist to make, in differ-ent contexts, two very different claims. In one grant proposal, they pointed out the barrenness of the straightened creek and asked for funding to correct it; in a second grant, they pointed to the water-monitoring device in the front to show that environmentalists were already active in the community working to make the creek a healthier place. That is, the particular story chosen frames what and how readers have to gaze at the photograph to see what its producer intends to communicate. In the case of the orchid, however, we should take into account the fact that in the bottom-right corner of this photograph, the word 'orchid' could be read. This detail could be regarded as a semiotic resource for identify-ing something relevant in the photograph, but it is certainly not enough to justify students' comments about the association of orchid and host plant.

Figure 6.5. A similar photograph at this spot was taken by environmentalist to make, in different contexts, two very different claims. In a grant proposal, they pointed out the barrenness of the straightened creek and asked for funding to correct it; in a second grant, they pointed to the water-monitoring device in the front to show that environmentalists were already active in the community working to make the creek a healthier place. (© Wolff-Michael Roth, used with permission.)

Captions

After pointing out several aspects of the photographs, the students claimed there was nothing else to be commented on. At this time, then, the interviewer presented the students with the captions of the photographs. These constitute additional semiotic resources that might alter how the students perceive and therefore interpret the photographs (given the relation is actually perceived, which is an empirical matter even though it is made highly probable because of the fact that they were introduced by the interviewer).

After reading the caption of the photograph of the orchid, which stated only 'Fig. 83.1 Epiphyte plant' (figure 6.1), all the twelve students worked to find the epiphyte plant in the photograph. The majority of the students (nine) correctly, even if tentatively, identified the epiphyte plant in the photograph. Three of these students actually referred to the orchid as being the epiphyte plant, and six students, although they could not say that the plant was an orchid, used deictic gestures (pointing) to identify the epiphyte plant in the photograph. Besides two students who already knew orchids from nature, seven other students proposed the orchid as the epiphyte plant in the photograph because it was the most focused or centralized object in the photograph: 'Because it is the most visible one, the one that is more like, like in the middle' (Ruth, eleventh grade), 'Because that's what more focused, together with the tree' (Andy, ninth grade), 'I don't know, it's showing them a lot' (Fran, ninth grade), and 'because it is the one that is more focused, more visualized I think, more apparent' (Adam, eleventh grade).

Reading the caption, the students focused their attention to a specific detail in the photograph, in this case, a plant. Lacking further resources in the caption that could help them to identify the epiphyte plant on this photograph, the students relied on other semiotic resources, made available directly on the photograph itself, such as, for example, differences in focus and alignment of the objects depicted.

In chapter 4, we classify this photograph as illustrative, because its caption provides only a name for the object or phenomenon the photograph is about. There is little else in the caption that might have helped readers identify this object or phenomenon. Therefore, it does not surprise that some students were confused about which plant was the epiphyte plant in the photograph. For two students, the epiphyte plant could be either the central tree or the orchid, and another student thought the epiphyte plant was the fungus on the branches of the central tree. This confusion is justified, insofar as the photograph lends itself as an illustration of lichens or another plants as much as it does as an illustration of an epiphyte plant. The presence of the caption referring to 'epiphyte plants' appears to exclude the possibility of the topic of this photograph being other plants than the orchid. However, considering that the students did not know what an epiphyte plant was, anything recognized as a plant in the photograph could be the object that this caption was referring to, including lichens.

Although the caption does not provide the readers with further resources that could help them to efficiently identify the epiphyte plant it refers to, most students pointed to the true referent of 'epiphyte plant' in the photograph. The caption allowed these students to separate and evaluate the gratuitous detail, which carries no relevant information, attribute it to the background, and consequently disregard it as irrelevant. In doing so, the relevant details actually become salient foreground, the real topic of the photograph.

Andy (ninth grade) was a good example of how this procedure was enacted. He identified the photograph as representing a plantation. When asked about what was the topic of the photograph, he said it referred to 'the part of the trunk of the tree' while pointing to the central tree. He justified this by saying, 'it's more concentrated here', while outlining the trunk of the central tree in the photograph. He further explained that if the photograph was about the plantation, it should have focused also on all the other plants in the background. After reading the caption, Andy identified the leaves (orchid) on the side of the trunk of the central tree as the epiphyte plant, and then he proceeded to analyze the photograph: 'if you analyze, really, there is only the trunk of the tree like there isn't its top, but then there is here this more focused'. In this example, the student actually engaged in the process of analyzing the photograph to separate background and foreground (figure), and to distinguish between the, perhaps gratuitous details and the intended topic of the photograph. Andy focused on the objects in the center of the image, which were in sharper focus, that is, the central tree and the leaves attached to it. He further refined his analysis by pointing out the absence of the tree top, which led him to perceive the tree as a secondary rather than the primary object of the photograph; rather, the leaves attached to its trunk was the more likely topic. The fact that the student regarded the orchid as something alien to the tree and not as 'the leaves *of* the tree' was relevant in his work of interpretation of this photograph. He separated 'tree' and 'leaves' and then decided which one was more likely to be the topic of the photograph, making use of other visual information displayed in the photograph to do just that.

In the case of the multiple photographs (camouflage), however, students' reactions after reading the caption were rather different than in the previous situation. Before reading the caption, only two students described the three photographs as depicting a sequence; five students perceived the birds to be of different species. After reading the caption, all the twelve students saw the three photographs as constituting a sequence of seasons or climates and that the bird represented in each photograph was the same animal. The most salient difference in the students' answers after they have read the caption, however, was related to the phenomenon of camouflage itself. Once the students had read the caption, they pointed out the fact that the white bird in the first photograph of the series was camouflaged, and it was almost impossible to see the bird in the photograph except for its beak and eyes. The same occurred with the last photograph in the series, where the students identified the phenomenon of camouflage. The fact that all the twelve students agreed upon these aspects of the series of photographs reinforces the idea that the complementary caption is preferable than the illustrative or explanatory captions, insofar as complementary captions helped identifying those elements that are required for understanding the representation in the way it was intended. The caption not merely described what was in the photographs but in fact taught the viewers what to identify as relevant detail. The caption enabled students to notice the relevant aspect of these photographs, namely, that a white bird blends with a similar environment during wintertime, and that, by changing its plumage, it also blends with the dry vegetation during summer time.

This particular series of photographs and its associated caption, however, also reveal a problematic dimension. The photographs and caption provide not only instances of a bird camouflaged against the environment, but it also presents a temporal sequence, on which the bird changes plumage according to the seasons, as the environment also changes. After reading the caption, three students described the series as presenting camouflage, half of the students referred to the changes the bird or the environment undergo, and three students suggested the series represented both events. An important part of the interpretation was centered on the second photograph; three students even said that the bird in this second photograph was not well camouflaged. Five students pointed out that this second photograph was in fact representing the change the animal goes through from how it is in the first photograph to how it presents itself in the third photograph, which lends itself to infer that these students identified at least the temporal-sequential nature of this series of photographs.

The students did not make sense of the second photograph in the context of the concept of camouflage presented by the caption. The students demonstrated that they understood the phenomenon of camouflage (two students even mentioned military strategies during a war and how the soldiers camouflaged themselves as examples of camouflage), and they identified the phenomenon of camouflage in the first and in the third photographs of the series, where the bird was very well camouflaged in its environment. They emphasized that the phenomenon of camouflage was very well illustrated in these two photographs; but they also pointed out that the photograph in the middle was not at all necessary. This second photograph actually interferes with the development of the camouflage concept, because it is a counter example. The caption further contributes to the confusion of camouflage and seasonal changes insofar as, in describing the photographs, it emphasizes these changes, while also stating that 'This is a good example of camouflage'. The caption refers at the same time to both events of camouflage and seasonal changes, without however helping the reader to identify the two different events represented in these photographs.

Main texts

After students had completed their interpretations of photographs and captions, they were presented with the main text associated with each photograph. Each text constitutes another set of semiotic resources that students can draw on to make sense of the photographs—whether they in fact do is an empirical question. In the present situation, the introduction of the main texts led to further changes in the students' interpretation. For example, after having completed the main text with which the orchid photograph was associated, all twelve students recognized the topic of the figure as being related to the orchid. They were, then, able to identify the leaves on the central tree as belonging to the orchid, although one student still demonstrated to have doubts about it, 'I don't know what is the orchid, what is the bromeliad but, there are plants here in the trunk of the trees, here there is one, see?' (Fran, ninth grade).

In the case of the camouflage series, the main text is about camouflage, explains the concept, and provides several examples of camouflage. However, the main text does not refer to the seasonal changes undergone by the bird. After reading the main text, ten students thought the topic of the photographs was camouflage but two students said the topic of these photographs was the seasonal changes on the plumage of the bird: '[These pictures] it's about European partridges, showing them changing their plumage in different seasons' (Fran, ninth grade) and 'It's about what place each species inhabits? . . . When it changes the season, the same bird starts to change the color of the feathers' (Ruth, ninth grade).

The main text and the caption cannot assure that students will actually interpret the photograph in the intended way. Alternative readings of the text can occur, as for example, in the case of a student (Gil, ninth grade) who identified the orchid as being the epiphyte plant, and the host tree as being a bromeliad. In this case, the alternative reading of the text was easily identified, as the student read 'orchids and bromeliads' instead of 'orchids or bromeliads' in the sentence, 'It is the case of the interaction between orchids or bromeliads and the trees in which trunk they are installed'. This alternative reading of the text (and instead of or) accounts for the difference in interpretation of the photograph and even of the entire concept presented in the text, insofar as the student perceived bromeliads as the host trees instead as another plant in the same category as orchids: 'the orchid is in a bromeliad, without causing any damage to it'.

In this particular case, the textbook author used the photograph to illustrate the *interaction* between orchids or bromeliads and the host trees (figure 6.1). Yet this interaction is not visually available in the photograph. Rather, it shows how an orchid looks like on the trunk of a host tree in a natural setting. The way in which this information was provided in the main text accounts for the possibility of different interpretations of the photograph. The text actually does not make explicit weather the photograph presents an orchid or a bromeliad; it only states that these are examples of epiphyte plants. Although the student in this case read the text in an alternative way, complete confusion would have been avoided or at least minimized if the text (or the caption for that matter) had explicitly instructed the readers that what they could see in the photograph was an orchid, and that orchids, as well as bromeliads, were epiphyte plants.

The work of reading the text and photograph simultaneously

'If you are interested in the picture, you will want to read what is beside it, or below it, or above it'. (Cameron, eighth grade)

Having presented the students with the entire assemblage of photograph and texts allows us to investigate graphicacy in practice, that is, the actual work of reading that the students engaged in when reading textbooks with illustration such as photographs. In these cases, we introduced students to the entire assemblage of photograph and texts together: we reproduced the page of the textbook where the photographs appeared and handed it to the students. The students were then asked to talk us through their reading of the page.

Common functions of photographs

One of the aspects that became salient in our analysis was the primary function that every photograph has in calling readers' attention. In the case of the photograph of the caterpillar, this function is not only the primary function as it is the only function of this photograph in the particular page where it originally appears in the textbook, insofar as this photograph had been previously classified as decorative, meaning that it does not present a caption or reference in the text associated with it.

As they received the reproduced pages, ten students immediately said that they were looking at the photograph; only two students stated that they were reading the title of the text or 'looking at the text'. In the case of the lichen photograph, it is easily noticeable from the videotapes that all the twelve students looked at the photograph before either talking about it or starting reading the text. The students' attention was immediately caught by the illustrations (only five students actually stated that they were looking at the photograph). As Cameron (eighth grade) stated, 'you look at an illustration, you will want to see what it is, if you are interested in the picture, you will want to read what is beside it, or below it, or above it'. If the texts do help the students in identifying what the figure is about, the pedagogical role of photographs is increased.

Contextualization in the work of interpretive graphicacy

Another aspect that emerged from our analysis is related to the importance of the context of the photographs for students' work of interpretation. Concerning the caterpillar photograph, five students said the photograph presented 'a caterpillar'; three students said it was 'a centipede', and two students referred to the animal just as 'a bug'. One student said it was 'a worm', and another one identified it as 'an insect'. Although the term used varied, all students identified the animal in the photograph. However, other aspects were also identified in this photograph: two students made references to where or when this photograph was took, 'in a closed space' (Andy, ninth grade) or 'at night' (Cameron, eighth grade), because the background was dark. These two students also inferred other aspects of the photograph. Cameron said, 'this is a small bug', and Andy stated that, 'it must stop to be able to eat'. We notice in these cases the influence of previous knowledge and conventions of perspective in the interpretation of photographs.

In the case of the photograph of the lichens, however, students' reactions were different: only three students commented about what they were seeing in the photograph, while other two students who did not start reading immediately said they were looking at the picture, without any further comment about it. The other seven students, although they clearly looked at the photograph (as can be seen on the videotapes), they did not make any comment and decided to start reading the text instead. This suggests that the students (with the exception of those three who actually tried to describe the photograph) did not identify anything in the photograph, and therefore turned their attention to the text, searching for information that could help them to figure out what the photograph was all about. A small number of students identified the background of this photograph either as a tree or a rock. Insofar as this information could not be found in the texts, the students relied on previous knowledge to identify the background of the photograph. The identification of the background is important because it allows the reader to distinguish the relevant details in the photographs. When this identification is not possible, the (by the author) desired interpretation of the photograph is jeopardized: there is a gulf between productive and interpretive graphicacy.

Comparing this situation with the one presented by the caterpillar photograph, we note that in the latter case, students were immediately drawn to the photograph. Although they all read the text, they all did so after commenting on the photograph. This makes it plausible to assume that in the case of the lichen photograph, students did not know what it was about. The visual information provided was insufficient to help the students interpret the photograph, and the texts associated with this photograph not only were important for directing the readers towards a specific interpretation, but also were essential to the reader to construct an understanding of this photograph.

The analysis of students' responses also reveals evidence of the influence of common sense and life experiences on the interpretation of photographs. For instance, when commenting on the caterpillar photograph, almost all students said the animal in the photograph was eating, feeding, or gnawing a plant or a leaf. Because the photograph is static, 'eating' has to be inferred. While 'eating' itself cannot be seen, possible evidence for it can be detected. Such evidence is built upon the difference between what one can see and what one may think, has heard, or believes. Here, the shape of the leaves is different than the way in which students knew it from experience; this difference can be hypothesized to be bite marks, and the caterpillar can be assumed to be responsible for these bites. A lot of previous knowledge and common sense, as well as conventions of perspective, goes into the work of interpreting this photograph in this manner. We assume, for instance, that caterpillars eat plants, or at least, leaves; therefore, these animals should have a mouth, and this mouth should be able to imprint a particular kind of mark in the leaves. All these details cannot be seen in the photograph and yet they are crucial aspects of the work of interpretation.

Indexical reference and photographs

High school biology textbooks inconsistently use indexical references—such as 'Fig. 30.2'—that link a particular segment of text (sentence, paragraph) and photograph; some photographs are not linked to the text at all. The orchid and the lichen photographs were accompanied by texts that presented indexical references to the figures, whereas the caterpillar and camouflage photographs were not referred to in the text.

There was no clear effect associated with the presence or absence of an indexical reference. In the case of the camouflage photographs, the absence of an indexical reference did not appear to be problematic at all, perhaps due to the distribution of text and figure in the page of the textbook. The text was placed in two columns and the figure was placed immediately below these columns. This arrangement allowed the readers to move from text to photograph in a continuous manner. In this case proximity accomplished what otherwise could only be accomplished with the presence of an indexical reference. The effect was different for the decorative photograph, here exemplified by the caterpillar photograph. In this situation, the lack of indexical reference aggravates the difficulty of associating photograph and text, particularly given the further absence of a caption. For instance, the text near the caterpillar photograph introduced the chapter topic; it was structured according to a typical pattern of the textbook in which it appeared: the first page of each new chapter presents a decorative photograph and an introductory text, as well as some highlights of the topics that would be presented in the chapter. While reading this text, four students identified the text as a summary or an introduction of what would be presented in the chapter that followed, but one that could be substituted by another picture:

I think [the picture] could be substituted by any other environment, insofar as it is the animal's environment, you couldn't put a horse in the sea, it wouldn't be right. But a horse in the field, it could be. (Adam, eleventh grade)

If they show a man there or even a bigger animal with a smaller one by its side and with a smaller one yet, then it would show that the stronger would... [this picture] here is good but you could use another too, maybe there are some other that would show it better. (Carol, eleventh grade)

The majority of students failed to identify this characteristic of the text and therefore were faced with increased difficulty in determining the function of the photograph in that particular text: 'What [the picture] is aiming to show, the meaning of it, I don't think I understand' (Brian, eleventh grade) and 'I think that it doesn't have anything to do with the text' (Faith, tenth grade).

Despite the difficulties, and the lack of caption and index, all students suggested that photograph and text were associated in some way, probably relative to living beings and the environment (students also mentioned ecosystem and biosphere). This relation, however, is a very general one and could refer to every single figure and text in this biology textbook. When asked about a more specific relation, there were as many different answers as there were students interviewed. That is, the students said the photograph was illustrating or exemplifying something referred to in the text, and each student believed this was a different thing: the natural cycles; the distinction between autotrophs and heterotrophs; the food chains; the particular environment (or ecosystem) where the caterpillar lives; the relation between the plant and the caterpillar; the importance of eating to surviving; the relation of human beings and the animals (and the fact that we should respect the animals); and metamorphosis. These were cited by the students as possible topics for the photograph, as an attempt to link photograph and text.

We can understand what happened in this situation in terms of a process of authentication. In this process, a photograph is viewed as evidence supporting the text, in fact, because of its lifelike quality, providing an image for what the text describes in words. In relating the text and photograph, the students at-

tempted to find a specific function to the photograph, that of illustrating something referred to in the text. They assumed the photograph was helpful in some
way to understand the text, and they struggled to justify this assumption by directly connecting text and photograph, even if this connection was not explicitly
available. The lack of resources that definitely and directly linked photograph
and text therefore gave rise to a level of indeterminacy that accounts for the
great variety of topics students proposed as the content of the photograph.

When a definite and direct link is provided, the level of indeterminacy
should be decreased. Thus, while reading the text associated with the lichen, six
students actually referred to the figure at the exact point in the text where the indexical reference was placed. Not all of them read the indexical reference aloud
at this point, but they either looked at the photograph, or pointed to it, or yet said
something about the figure. The students were aware that the photograph represented lichens; both caption and main text read that the photograph was about lichens, and the indexical reference appropriately placed in the text contributed to
help the students to identify the right detail, that is, the topic of this photograph.

Some students even stated that they already knew what lichens look like,
and that they were able to identify it in the photograph before reading the caption and the main text (although only two students actually named what they
were seeing in the photograph before reading the text). When asked to point out
the lichens in the photograph, eight students were able to do it properly, that is,
identifying the lichens against the remaining background. However, they were
not sure about what was the background constituted of.

In the case of the photograph of the orchid, the presence of the indexical
reference to the figure in the main text played an important role in guiding the
students towards the interpretation of the photograph, which emerged from the
association of this photograph with its caption and text. The indexical reference
helped students in selecting the specific topic of the photograph in relation to the
main text. In this case, for instance, the presence of the indexical reference just
after the phrase, 'It is the case of the interaction between orchids or bromeliads
and the trees in which trunk they are installed', leads the readers to regard the
photograph as providing an example of the interaction referred to in this phrase.
Readers, then, are expected to look for a plant like an orchid or bromeliad and
the trunk where they should be installed.

Although not every student read the indexical reference in the main text
aloud, the majority of the students recognized the presence of an indexical reference that linked figure and text. The others either did not mention the indexical
reference or they stated that there was no indexical reference to the photograph
in the main text (one student even read the indexical reference aloud when read-

ing the main text, but he failed to recollect its existence when asked about it by the interviewer).

Although the index to the photograph of the orchid is important—especially because this photograph is an illustrative photograph that does not present much useful information in the caption—the fact that the indexical reference is placed at the end of an entire phrase in the text accounts for an unexpected difficulty in associating photograph and text. For instance, Fran (eleventh grade) engaged in an effort of identifying *both* orchid *and* bromeliad in the photograph. She could not decide whether the leaves on the side of the trunk of the central tree in the photograph were an orchid or a bromeliad. She was also confused when she attempted to identify a second, epiphyte plant in the photograph thus pointing to many different trees and flowers in the photograph.

The way in which the indexical reference was used, accounts for the possibility of different interpretations of the photograph, as for example, in the case mentioned above. The reader is taken to connect figure and text through the indexical reference placed at the end of the phrase where the text refers to orchids *and* bromeliads. Insofar as in this case the caption does not help to disambiguate matters, the student could not decide if the epiphyte plant in the photograph is an orchid or a bromeliad.

Beyond text, caption, and indexical reference

Visual information such as color, arrows, letters, geometric forms, and so on could be used in addition to caption and text to highlight something directly on the photograph. However, as we showed in chapter 4, high school biology textbook authors rarely use such semiotic resources with photographs. One example of the use of an additional semiotic resource layered on top of the photograph existed in the orchid photograph. In this situation, the information necessary for the reader to reconstruct the topic of the photograph (orchid or bromeliad) was available in the photograph itself, existing in the inscription 'orchid' on the bottom right. Taken in isolation, this information does not help the students to actually identify this specific object in the photograph, although it could still focus the readers' interest towards something specific in the photograph. But in the context of the entire assemblage of photograph, caption, and main text including the indexical reference, this information becomes essential for the reader to make sense of the photograph in relation to the text. Just by looking at the photograph, the students identified the most central objects as the probable topic of the photograph; by reading the caption, they focused their attention at a plant in the photograph, and later, by reading the text, they associated the figure with the text, going as far as realizing that the photograph was showing an epiphyte plant and its host tree. However, they remained uncertain about which of the examples of epiphyte plants given by the text—orchid and bromeliad—was represented in the photograph. The word 'orchid' inscribed into the photograph was the ultimate information needed to properly interpret this photograph in relation to the text. Therefore, photographs, captions, texts, indices, and a variety of other resources can and should be used to make sense of photographs when reading a textbook. The work of interpretation of photograph and text is essentially dialectic. As the student quoted at the beginning of this section noted, the text helps to understand the photograph as much as the photograph helps to understand the text, and both text and photograph need each other in order to be properly interpreted.

Role of interview and previous knowledge and experiences

'I thought, "it's a research, so there will be some trick in here"'. (Adam, eleventh grade)

Seldom addressed in the literature but nevertheless an important semiotic resource to interview participants is the interview context itself as well as every word, sentence, and even pause produced by the interviewer.[4] That is, interpretive graphicacy is mediated not only by the prior experiences that individuals bring to and with which they elaborate an inscription, but also by the very interview setting, which provides interpretive resources that are rallied by completing the task at hand. We observed such influences during the present interviews as well. For example, while looking at the photograph of the orchid, Adam (eleventh grade) was asked to articulate the topic of the photograph. He immediately answered that it would be the orchid, and he pointed out the orchid in the photograph. He also pointed out the word 'orchid' written in the bottom right of the photograph. After reading the caption, however, Adam suddenly changed his mind. He said the plant he earlier had identified as an orchid no longer looked like an orchid, and he added that there are many kinds of orchids and the ones he knew were different from the plant in the photograph. He said he was confused and that the caption only makes everything worse. Only after reading the main text Adam returned to talking about it as an orchid and an epiphyte plant. He realized then that both denominations could be used to address the same plant in the photograph.

During the debriefing session, Adam admitted that he knew the plant was an orchid from the beginning, he had recognized it in the photograph because he was very familiar with orchids. His mother grows orchids at home and his grandmother's garden includes many orchids growing on trees. However, when he read the caption, although he did not know what an epiphyte plant was, he believed the word 'orchid' in the bottom right of the photograph was deceptive and did not have anything to do with the photograph itself. He considered the plant identified as an orchid as being the epiphyte plant referred to in the caption. He explained that because he knew the interview was part of a research, he expected some kind of trick that would mislead him.

In this case the influence of the context in the answers provided by the student is clear. Although the student answered the interviewer's questions, he was also conscious of the fact that the interview was part of a research project, and his conception of the nature of research influenced his responses to the extent that he disregarded his own previous, extensive knowledge about orchids.

This chapter was designed such that we could investigate differences related to students' previous knowledge when reading the photographs and texts. Thus, half of the students had already had an advanced ecology course in high school, and the other half had not. We expected to be able to distinguish differences on students' responses from those two distinct groups. During our analysis, no differences on students' responses were salient enough to justify being mentioned in this study. However, we were left with the overall 'feeling' that the older and more schooled students from eleventh grade did demonstrate some advantage in relation to the younger eighth- and ninth-grade students. This assumption is based on the fact that, overall, eleventh-grade students were faster in associating information provided on captions and main texts with the photographs. For instance, although most of the students, from both groups, encountered difficulties when first trying to interpret the photographs, eleventh-grade students more easily changed their interpretations after they had read the captions or texts accompanying the photographs. The older students also answered the questions posed by the interviewer in a more matter-of-fact manner, whereas the eighth- and ninth-grades students hesitated more when answering the same question, and demonstrated to have more doubts. We think that older students, who have been attending school for longer, did develop some kind of literacy concerning reading textbooks with pictures, even if this literacy was not explicitly taught at school, as part of the curriculum.

Autonomously developed graphicacy

With the study presented here, we wanted to find out what levels of graphicacy that those students display who did not have particular instruction to develop the knowledgeability with respect to graphicacy. That is, our interviews elicited levels of graphicacy (Brazilian) students develop as part of their everyday lives in and out-of school. In a sense, ours is baseline work that should help interested teachers, curriculum designers, and researchers to design learning environments that allow students to develop graphicacy and its reflexive component, *critical graphicacy*.

We began this project knowing that teaching and learning strategies rely heavily on textbooks. We were therefore interested in carefully investigating (a) the pedagogical potential of photographs and (b) how students make use of these visual resources to achieve and help others to achieve understanding. Our analysis reveals differences between single and multiple photographs. External comparison provided by the use of series of photographs allows the readers to easily distinguish differences and similarities between photographs. Thus, when interpreting the photographs of camouflage, students could easily identify the differences on the plumage of the bird and on the environment. It was not so obvious to them, however, that the first and the third photographs of the series constituted examples of camouflage. Internal comparisons of each individual photograph became secondary to the work of interpretation. In these situations, other resources (caption, texts) were employed, when present, for highlighting important aspects to be observed, and guide the readers through their work of interpretation. Therefore, a series or a pair of photographs by itself is not enough to ensure the correct interpretation of the photographs. It is the interaction of all semiotic resources presented in the textbook, together with the photographs, that made possible to readers to interpret and understand what they are reading. In the case of single photographs, even though the internal comparisons are immediately fostered, external resources still need to be associated to the photograph itself, during students' working of reading these photographs.

The caption is a major aspect when using photographs for pedagogical purposes in textbooks. Captions name something that should be more carefully regarded in the photograph. Decorative photographs, without captions, proved to generate greater difficulty in associating photograph and text, as this association becomes subjective when explicit links (such as caption and indexical reference, for example) are missing. Therefore, it is important that every single photograph has its own caption when it is used for pedagogic purposes.

When the caption fails to help readers to unmistakably identify the 'right detail' in the photograph, its existence is no longer essential to readers' work of interpretation, insofar as they have to draw on resources other than the caption to identify the topic of the photograph. For instance, most students successfully identified the orchid in the photograph by relying in characteristics inherent to the photograph itself, such as, for example, framing and focusing. These characteristics then became semiotic resources for the students to understand the photograph when the caption fails to help them to do so.

Our investigation reveals that main texts are certainly an important resource in helping the readers to interpret the photographs. Complete explanations about the object or phenomenon depicted in the photograph, as well as appropriate associations between the concept been presented and what can be seen in the photograph are necessary to maximize the chances that readers will, in fact, connect and interpret text and photograph in the way expected by the textbook author.

Our analysis makes it quite clear that one of the major functions of photographs is to capture readers' attention. The students interviewed noticed and commented on the photographs before referring to the texts. Although we were careful as to take into account the influence of the interview setting as well as the influence of the interviewer herself on student's responses, we do believe the students demonstrated interest in the photographs, even if they did not pursue the investigation of these photographs afterwards.

It is important to be attentive to intrinsic characteristics of photographs, as for example the background and framing of the photographs. Photographs with neutral backgrounds are useful to highlight the object of interest, insofar as gratuitous details are almost non-existent. Nonetheless, the lack of these gratuitous details sometimes implies in the decontextualization of the object depicted in the photograph. In the lichen photograph, for example, we could notice how the students could not comment about the photograph before reading the text, as they could not easily identify the object in this photograph. The photograph was taken aiming to focus only on the lichens, and although these were fairly represented in the picture, they were rendered out of context, which generated difficulties for the students to understand this photograph. Furthermore, the information in the texts accompanying this photograph was not enough to help the readers to distinguish what they were seeing in the background.

On the other hand, we show here that some photographs, as that of the orchid, also present students' struggles related to the identification of the important or right detail to be observed in the photograph. That is, our study highlights the gap between the productive graphicacy of the photographer and textbook author and the interpretive graphicacy brought to the photographs by the students. The recovery of the author-intended sense and sense of photographs requires corresponding levels of graphicacy; deconstructing the sense and reference conveyed requires an additional reflexive turn, critical graphicacy. Many students do not display such levels; their autonomously developed knowledgeable graphicacy does not allow them to arrive at the intended senses, references, and meanings. The abundance of details in this photograph certainly gives it realism; unfortunately, it also accounts for much confusion. Students do not unmistakably identify the epiphyte plant among so many other plants depicted in the photograph. Some students do not arrive at a decision about which plant was the epiphyte plant even after reading the caption and the main text. We again suggest that more specific directions, such as arrows and colored areas, should be used to help the readers to identify the right detail to be observed in the photograph.

Our analyses show that students do in fact pay attention to the indexical references to the figure when reading a textbook. The indexical reference to the lichens photograph provided a good example. When reading the text, the students immediately connected 'lichens' with the photograph, because the index for this photograph is placed just after the word 'lichens' in the text. In the case of the photograph of the orchid, however, the indexical reference placed after an entire phrase that identifies a phenomenon and not an object, confuses the reader. The indexical reference is what allows the reader to connect photograph and text; therefore, it can be an essential resource to help readers to interpret photographs and texts in the context of learning a scientific concept.

7 *Layered* inscriptions: what does it take to get their point?

Our close analyses of inscriptions in the previous chapters showed that reading inscriptions such as graphs or photographs and integrating them with textual information requires a tremendous, perhaps insurmountable amount of work. On the one hand, students may not have had the resources for doing such work; on the other hand, textbook authors may not have done enough to facilitate doing this work. How might textbook authors facilitate students' efforts in reading inscriptions, particularly those where multiple inscriptions are drawn on top of one another? A hint for how this might be achieved comes from research on the use of inscriptions in computing environments.[1] These studies showed that inscriptions of different types presented simultaneously on top of one another, such as simulated objects and vectors that represent force and velocity, assisted the learning of kinematics.

Assuming that the inscriptions used in textbooks have the purpose of teaching *specific* concepts, one is tempted to ask, 'How might the layering of inscriptions provide students with resources in learning science from textbooks?' and 'Why might any such mediation occur?' For example, one might ask, 'What is the work of reading required to understand an inscription that layers a graph displaying Boyle's law, naturalistic renderings of pistons, and force arrows?' (figure 7.1); and 'What does the layering do that simple forms of inscriptions do not achieve?' We begin by answering the question, 'What work is required for reading these inscriptions to get their point, that is, to learn the *specific* concept(s) that this inscription is intended to teach?' Answers to all of these questions provide us with a better understanding of the nature of knowledgeable graphicacy with respect to inscriptions that consist of multiple layers.

The text that accompanies figure 7.1 indicates that the figure represents the relationship between the volume and pressure of a gas, that is, Boyle's law. Boyle's law is articulated in terms of the statement 'at the same temperature, the volume of some gas is inversely related to pressure' and the equation '$P \times V = k$'.

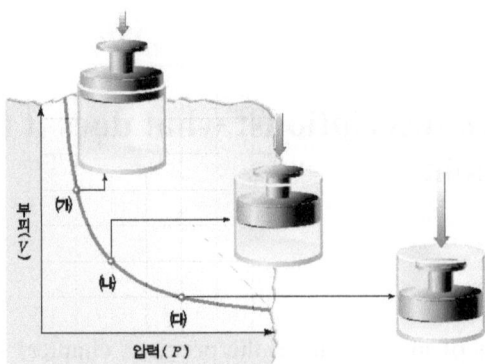

Figure 7.1. This example of a layered inscription was taken from a Korean grade-seven textbook in the section on Boyle's law. The label on the ordinate is 'volume (V)'; the label on the abscissa is 'pressure (P)'. The letters on the graph are 'a', 'b', and 'c', respectively, from top left to bottom right. (Reprinted with permission of the copyright holder)

The figure itself presents a graph, seemingly torn from a textbook, superposed by two different types of inscriptions. First, there are naturalistic drawings rendering grey weights (or pistons) in a green but apparently transparent beaker. Second, there are three yellow and orange arrows of different length positioned above each beaker-piston combination. Fine arrows in black begin at different positions on the graph and each point to one of the three beakers. The stated purpose of the inscription is to allow students to learn Boyle's law (as in the mathematical inscription P × V = const), embodied in the graph; not stated is the fact that students need to ground the inscription in their lived experience and understanding of how the world works. What is the work of reading required to relate this layered inscription to one's lived experience, and therefore to learn from reading or interpreting this inscription? How does this work differ from other circumstances that either state Boyle's law simply in its mathematical or in mathematical and graphical form?

At a global level, the inscription was designed to mediate students' learning of Boyle's law, normally stated in the form of 'P × V = *k*', and often expressed in terms of a graph. The beaker-piston combinations and yellow-orange arrows are additional resources that potentially mediate between the more experience-distant equation and graph and the experientially nearer beaker-piston combination. Although this inscription might look easy to the (science) educator and (science) teacher who already knows it and knows about Boyle's law, it is rather complex work that is hidden and needs to be done.

Our initial textual presentation of the inscription already articulates the first type of work to be done. That is, at a global level, readers[2] have to perceive the three types of inscriptions as *separated* yet connected inscriptions, constituted at the microlevel by colored dots (print) on the page: the red graph on light-blue lined paper, grey pistons in green beakers, and yellow-orange arrows. At a more fine-grained level, readers have to articulate, for example, the green areas as beakers and the grey areas as pistons or weights. Here already we encounter more work to be done: are these grey entities generic pistons or are they specific weights? More work is required, for example, in the form of comparing the three beaker-piston combinations, which is work within the same type of inscription. A comparison of the three pistons reveals that they are equal in size; they are, however, inserted into the beaker at different depths. In fact, for our reading to take us to Boyle's law, we need to *see*—i.e., do perceptual work—the amount of space left on the bottom of the beaker rather than how far the piston has descended into the beaker. That is, if the grey areas do represent weights, then our experiences would suggest that they are of the same weight—unless they were made of different materials (which requires experiences with and understanding of the density concept). Perceptual structuring further reveals that the three pistons are inserted into the beaker at different heights. *Comparison* (work) of the different heights with the sameness of the grey parts suggests the latter to be generic pistons (same weights) rather than different weights that would require special attention.

These are only drawings of beakers and pistons. Work is required to see the grey part of the drawing *as a* piston, the green parts of the drawing *as a* transparent beaker, that is, work is required to relate the drawings to corresponding things in the world we know so well. Even such apparently simple relations between a drawing of a thing and the thing that it denotes—iconic relations in the language of semiotics—are learned and culturally specific. These relations require previous experiences with such things as beakers and pistons (or weights) and with cultural conventions regulating the relationship between drawings and the things they depict.

There is more work involved. *Perceptual structuring* reveals that the color of the wide arrow changes from yellow at the tail to orange at the tip. Within-inscription-type *comparison* reveals that the three arrows have the same width but are different in length. Between-inscription-type *comparison* is required to produce the inverse relation between the length of the arrow, on the one hand, and the distance of the piston from the bottom of the beaker, on the other hand.

In chapter 1, we show that signs have to be perceived as such before they can be interpreted. For Sherlock Holmes, only some things he can see in a room point him to the murderer, other things are incidental; but at the outset of the story he does not know. In interpreting a graph, only some things that can be seen are relevant to the interpretation, others are incidental; for example, the birthrate and death rate graphs in chapter 1 require comparison between the two lines, not a comparison of their slopes, which we have seen even experienced scientists doing. Relative to the present inscription, perceptual structuring distinguishes the paper and grid (here blue) from the graph proper, here black axes and red line. The black lines are not just axes but, as indicated by the arrows, are ordinate and abscissa of a grid system where distance from the intersection is equivalent to magnitude. Thus, although not specified in the inscription, a relationship to algebra and numbers needs to be made. The red line has to be articulated (work) as part of this grid rather than of other parts of the inscription. It moves from top left to bottom right in a smooth curve. Each point has to be constructed (work) as a couplet relating a particular value of volume and pressure. (This also requires 'pressure' and 'volume' to be associated with abscissa and ordinate, respectively.)

On the red graph line, there are three blue circles; these require perceptual structuring as points of the red line. Black arrows are drawn from each of these points to one of the green beaker-piston combinations, two pointing to the sides, one to the bottom of a beaker. If the drawings are to assist in learning, the student needs to do multiple relation work: (a) for each point, relate the length of the yellow-orange arrow to the distance of a graph point from the origin along the ordinate, and relate the height of the piston above the beaker bottom to the distance of the point from the origin along the abscissa; (b) relate these relations to one another.

The black arrows do not link each member of a couplet (P, V) to the corresponding height above the bottom in the beaker and the yellow-orange arrow, but generically point to the beaker. This may increase the amount of relational work that has to be done by the reader. That is, the reader needs to perceive each point as having its own value of P and V, and link those to specific features of the beaker-piston combination and to the yellow-orange arrow. Careful comparison is required to link the P-values (distances of graph points from the origin along the ordinate) to the lengths of the yellow-orange arrows, and to link the V-values (distances of graph points from the origin along the abscissa) to the heights of pistons above the beaker bottom. The comparison among these links will reveal to the reader that there is an inverse relation between P-values and V-values, both in the graph points and in the arrows and beaker-piston drawings.

This inverse relation—Boyle's law—should be related to the reader's real experiences. Readers may, or may not, have pushed and pulled or released a piston inserted in a syringe (beaker) or have used a bicycle pump closing its end with a finger. If they pushed the piston more forcefully, the distance of the piston from the bottom of the syringe would decrease more, as the pressure on the piston would change the volume of the gas trapped inside of the syringe. The memory traces of these experiences could then be linked to the drawings of the piston-beaker combination and the accompanying three arrows, and to the graph representing Boyle's law. In this way, the drawings (arrows) could bridge the gap between the graph and the real world. The question is, however, what do readers relate the layered inscription to if they do not have this prior experience?

In this chapter we describe the reading work required to understand inscriptions in general and layered inscriptions in particular. We articulate the work in terms of a semantic model that makes salient the different types and amounts of work required to the interpretation of an inscription. We provide several analyses of different types of layered inscriptions to exemplify what science educators and textbook writers have to consider when preparing layered inscriptions with the idea of assisting students to get to the point of an inscription, that is, learn the specific lesson that the inscription was intended to teach.

Semantic model of multiple (layered) inscription

We began the construction of our semantic model[3] with existing studies concerning interpretation of inscriptions. The work of moving from one type of inscription to another type of inscription is called a *translation*; the work of moving from one to another inscription of the same type is called a *transposition*. Examples of translations include the construction of a graph from a table, or the interpretation of a graph, which requires its translation into a verbal description. A translation, therefore, corresponds to moving along the inscription continuum presented in figure 4.2; each move requires work because there is an ontological gap between any two neighboring inscriptions. Examples of transpositions include the creation of one graph from another graph or the creation of an ordered data table from an unordered one. Unnoticed by most educators is the fact that translation and transposition require the existence of equivalent structures— different natural phenomena are articulated whether the height or slope of a graph is salient to the interpreter. That is, to understand interpretation researchers also need to attend to the *perceptual structuring* of inscriptions (and natural

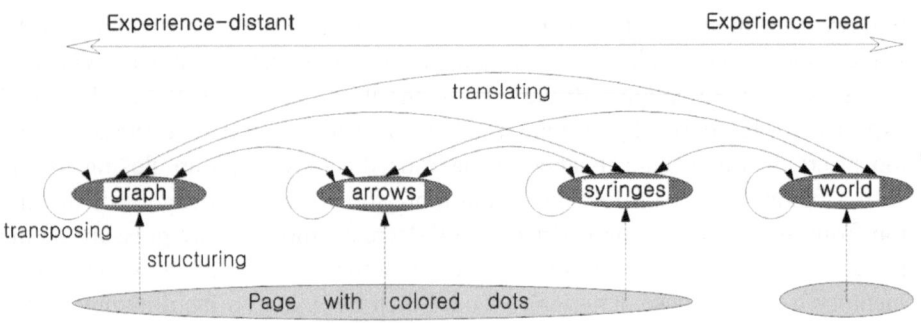

Figure 7.2. Semantic model of reading the layered inscription in Figure 7.1. Each arrow denotes work to be done in reading this inscription: structuring (dotted arrows), transposing (circular arrows), and translating (curved line arrows). The progression from left to right is from more experience-distant to more experience-near inscription, and experience (world) itself.

phenomena) from their raw material, the colored traces on paper or colored dots on computer monitors. These three types of work, structuring, transposing, and translating constitute the basic elements (types of work) of our semantic model (figure 7.2), which we exemplify by returning to our introductory inscription concerning Boyle's law (figure 7.1).

Inscriptions (graph, arrows, drawing) and the real world experience of the reader are represented by two ovals (figure 7.2). The lower oval (the elongated gray circle) refers to (a) the material basis of an inscription, that is, the colored ink dots that constitute figure 7.1 and (b) the unmediated, lived experience of the world. The upper oval refers to the structure that appears in the perception of the interpreter: on the one side, the structure corresponds to the signs that are interpreted, on the other side, the structure corresponds to the ways readers have come to *account* of their experience (for example, in their life narratives). To be able to make the required connections implied, this material basis has to be structured by the reader into a set of inscriptions with internal equivalent structures that can be linked—such equivalent structuring of educational materials on the part of the student cannot be taken for granted and often is not the case.[4] How these ink dots appear to readers depends on their structuring work. In the introductory example, the overall structuring into separate inscriptions is facilitated by the use of distinct colors associated with each inscription—blue for the paper, red for the graph, orange for the arrows, green for the cylinder, and grey for the piston. In our model (figure 7.2), the different structured domains (upper

level ovals) are ordered constituting a chain of references: from the most experience-near on the right, lived experience in the world characterized by its local,
particular, and continuous nature, to the most experience-distant on the left, here
graphs, characterized by increasing non-local, standardized, and universal character (see also chapters 3 and 4 on this point).

Arriving at a separation of the same material basis into the intended cascade
of inscriptions does not mean that students will arrive at the required internal
structuring of each inscription. In figure 7.1, for example, a person might perceive a piston or a weight; and the reader might perceive the piston as going a
certain distance into or as being a certain distance from the bottom of the cylinder. These are different ways of perceiving the same material configuration. In
each case, what the reader perceives has emerged as the result of structuring
work. Most frequently, structuring is automatic so that we do not attend to it.
But when readers of inscriptions experience problems, they may say to themselves something like, 'I have to look at it in a different way'; in this, they explicitly referred to the work of structuring the raw materials before them. There
is evidence from our own research among scientists that the structured ways in
which they come to look at a natural phenomenon and the mathematical structures that evolve as part of their research emerge jointly.

Perceiving the different piston-cylinder configurations as different states of
the same piston and cylinder ensemble requires the work of transposition (circular arrows in the model [figure 7.2]). This work is facilitated here because the
three different configurations are available and therefore can be more directly
compared than if the reader had to imagine such a change. A translation is required in going, for example, from the value of pressure at a particular point to
the length of a yellow-orange arrow. The work of translation occurs when a
reader makes a relation between the changing height of the three pistons above
the cylinder bottoms and the changing volumes on the graph. Another translation is involved when readers make links between the drawing of the pistons and
their real world experiences with pistons. Series of translations support an understanding of the relation between inscriptions and, ultimately, an understanding of the scientific content that all inscriptions refer to together. For example, in
figure 7.1, a person might come to understand the relation between pressure and
volume—Boyle's law—both within an inscription (e.g., graph) and between inscriptions (e.g., arrows and drawings) and the world (e.g., his experience of syringe with piston) they are standing for.

Our semantic model articulates the layering effect in two ways. First, there
is one material basis making all inscriptions appear in the same plane. Second,
when we look at figure 7.2 from the right side, all structured inscriptions come

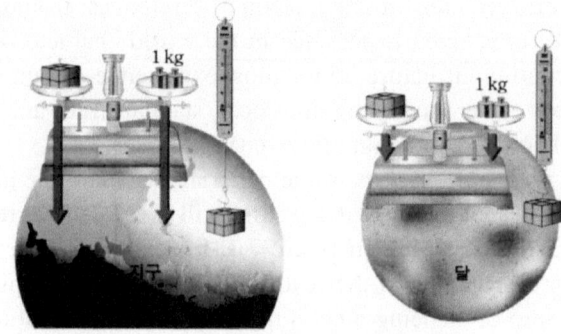

지구와 달에서 측정한 무게와 질량

Figure 7.3. Example of background layered inscriptions. The caption reads 'Weight and mass measured on the earth and on the moon' (the letters on the drawing reads 'the earth' and 'the moon', respectively, from left to right). (Reprinted with permission of the copyright holder.)

to be aligned so that between the real-world experience and the graph two other inscriptions appear, the drawing and the arrows. Here we focus on the role of the intervening inscriptions, which potentially play a mediating role, allowing readers to get to the point of inscriptions by bridging the gap between lived experience and graph.

The work of reading layered inscriptions

In this chapter, we develop descriptions of the work that readers have to do to relate experience-distant inscriptions to their own lived experience; that is, we show how lived experience becomes an integral part of graphicacy. We focus on the role of additional inscriptions layered on top of the target inscription. Our initial analyses made it clear that there were differences between layered inscriptions. Because there is a possibility that these differences would lead to different kinds and amounts of reading and interpretive work to be done, we constructed a classification scheme covering all 447 inscriptions that we have found in two Korean seventh-grade science textbooks. We arrived at eleven such functions, which we captured in the following category names: Simple (no apparent function), worksheet, background, analogy, data-presentation, magnification, sequencing, systemic relation, comparison, filter, and explanation. The eleven functions are defined as follows: (a) simple: two or more inscriptions are just layered together. There seems to be no important relationship between the inscriptions; (b) worksheet: two or more inscriptions where parts of them are to be filled in by the reader, thus becoming layered inscriptions. Empty spaces are provided where the reader should draw something; (c) background: two or more inscriptions where one of them serves as a background to the other, providing specific contexts; (d) analogical: two or more inscriptions where the relationship between inscriptions is analogical to one another; (e) data-presenting: photographs or drawings are inserted in a table as examples of phenomena; (f) magnification: one or more inscriptions show precisely the relevant part of the other inscription, through magnification—sometimes the magnified parts give the readers more detailed information, especially in biology; (g) sequence: each inscription represents one step or the result of an experiment or activity—sometimes, this takes the form of time-series presentations; (h) systemic relations: diagrams and systems that have arrows representing movement or causal relationship—diagrams depict the mechanism or the flows with arrows layered on top of another inscription whereas systems are causal models representing both the entities (natural objects or phenomena) and their relationships with arrows, lines, or words; (i) comparison: two or more inscriptions present phenomena to be compared to each other; (j) filter: photograph with a drawing that has undergone some filtering; (k) explanation: two or more inscriptions that explain the topic (contents, principles, models, and natural phenomena) together. Each functional relationship presents a different kind or amount of reader's work necessary to make sense of a layered inscription. In addition, the categories are not mutually exclusive, because we can find several cases of layered inscriptions that have two or three functional relationships. For example, one of the figures discussed below (figure 7.5) includes both analogical and explanatory layer.

We select four examples as representative of our sample, simultaneously showing one or more kinds of layered inscriptions. We choose these four inscriptions because they cross categories, including multiple layers, which allowed us to exemplify a maximum number of functions with the least number of examples. Thus, the following example of *background* layers also contains *magnification* and *comparison* dimensions; the section on *analogical* layers contains *explanatory* dimensions; and the example of *data-presenting* layers also contains *filter* and *comparison* dimensions. Before proceeding, however, we provide the following commentary on special inscriptions. Simple layered inscriptions are layered images that are combined together using resources, such as, for example, computer software programs that allow the superposition of two images. Although no other relationship between the layered inscriptions is evident, each inscription is somehow related to the text (at least evident to those readers who are familiar with the domain). The layering of these inscriptions does not produce or multiply additional meaning that would exist by layering inscriptions together. Therefore, this kind of layered inscription constitutes a sort of exception in our scheme because it does not require translation (relating work) between the constitutive inscriptions.

Many explanatory diagrams or schemas with arrows that represent the flow of materials through various organs are used in biology-related sections of textbooks (systemic relations). The organs are conventionally depicted in cross-sectional perspective, onto which arrows are layered to articulate the movement of, for example, nutrients. The arrows soften the static character of the image. This kind of layered inscription can be considered as being layered *inside* another inscription, whereas other categories of layered inscriptions present an inscription layered *outside* another.

Background layer

Background layered inscriptions include situations in which one inscription serves as a background to another, therefore providing specific contexts with which another inscription can be read. The content of the background inscription is related to the topic of the main inscription, which can be, for example, a table, pie chart, graph, drawing, and so on. An example of a background layer can be found in figure 7.3: readers are to learn from this inscription that the weights on these two celestial bodies are different, because of the difference in the gravitational forces of the earth and the moon, but that the masses are the same. To learn *this* lesson, the learner has to do very specific types and amounts of work, which we articulate in the following.

The work of reading figure 7.3 is modeled in figure 7.4. The inscription in figure 7.3 actually contains four different layers. The drawings of two circles are visible, with one of the circles cut horizontally at the bottom. The circles containing the inscription 'earth' and 'moon', respectively, serve as background to naturalistic drawings of equal-arm balances and spring scales. Blue arrows are layered onto each side of both balances, beginning at the arms immediately below the pans. The caption associated with the inscriptions reads 'weight and mass measured on the earth and on the moon'. The blue arrows and drawings of the two circles with their own names are designed to bridge the gap between real world (far right in figure 7.4) and the text (caption) (far left in figure 7.4) that are experience distant. For completeness, our model also includes, as a separate

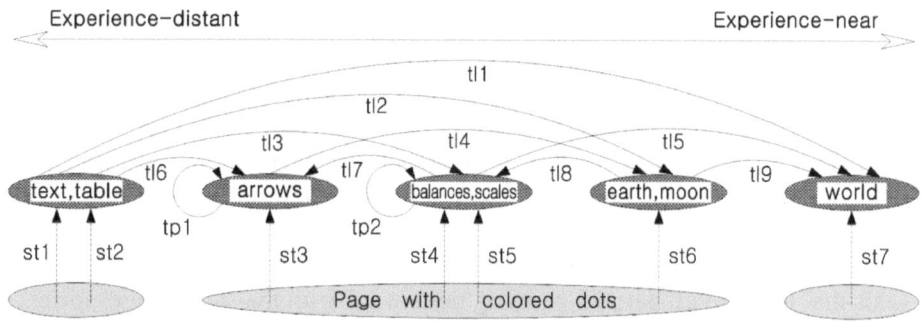

Figure 7.4. A representation of works required for reading the layered inscription in Figure 7.3. Each arrow represents the work of structuring ([st], dotted arrows), transposing ([tp], circular arrows), or translating ([tl], curved line arrows).

material base, the table featured elsewhere on the page in the textbook and not available in our reproduction of the inscription (figure 7.3).

To unpack this inscription and therefore arrive at what is pedagogically intended, initial perceptual work is required to differentiate the three different layers from the material basis of figure 7.3 (elongated circle in figure 7.4) followed by the work for structuring, transposing, and translating within and between the layers. The different types of inscriptions to be separated are the two circles (the left one has orange–white–blue–black colors, and the right one has grey color), the naturalistic drawings of balances and scales (with grey and orange colors), and the blue arrows (which would not be available in the real equal-arm balances). Two equal-arm balances and two spring scales need to be perceived on top of each circle (that is, translation tl8 [figure 7.4]). The equal-arm balances have to be articulated as weighing 'orange cubes' in comparison with two grey weights placed on each pan (structuring st4 [figure 7.4]). The same 'orange cubes' are hanging from the spring scales (st5). The two grey weights on the right pan of each equal-arm balance need to be understood as weighing 0.5 kilograms each, as it can be assumed that together they weight 1.0 kilogram (st4). All balances and scales must be assumed as being balanced (st4, st5), being in the final equilibrium state, based on, for example, the fact that the needles perpendicular to the beam of the equal-arm balances are not skewed to either side, providing, of course, that the balances are laid on a horizontal firm plane. For this assumption readers are required to recall previous experiences on weighing things with balances and scales (tl5).

Perceptual structuring further reveals that the 'orange cubes' are not just weights but also that they are all the *same* weight on either one of the two equal-arm balances and the two spring scales. This fact constitutes the basis for comparing weights and masses. Comparing scales (tp2 [figure 7.4]) on the left and right sides of figure 7.3 reveals that the equal-arm balances are balanced similarly, but the spring scale on the left side has a longer string between the weight (orange cube) and the body of the scale than the spring scale on the right side. Perceptual structuring of the blue arrows (within-inscription comparison tp1) reveals that they are all parallel pointing downward with the same length for each one of the two pans of the same equal-arm balance, but with different lengths between the left and the right equal-arm balances (st3). However, the texts do not state why the lengths of the blue arrows are different between two equal-arm balances, which would be a cue that leads to learn about different weights on the earth and the moon. There is also no explanation of what the arrows stand for, and, consequently, it is not evident why the arrows are layered in that position, just under the pans of the equal-arm balances. The arrows have the same blue color, which facilitates their perception as standing for the same or similar entities and events (st3). The function of the blue arrows has to be found while reading other (background) layered inscription, text, and their relationships.

The two drawings in the form of a circle have to be seen as depicting different things (st6). The circle on the left, which is cut horizontally on the bottom, has various colors and lines. Sharp boundaries of different colors have to be perceived as representing different things, that is, the south Asian continent and the sea (tl9 and st7): readers have to resort to experiences with terrestrial maps, whereas the contrasting colors of orange, white, blue, and black (which the instructed person knows to stand for the area of ocean) should be ignored, as they constitute structure irrelevant to the pedagogy of the inscription (st6). The words in the center of the left circle, 'the Earth', are consistent with the perception that this circle is in fact representing the planet Earth, with its continents and oceans. The circle on the right has to be perceived as depicting the moon, ignoring the strange gray color and texture of this circle, and considering the words 'the moon' written inside the circle (st6).

The information provided in the caption of these layered inscriptions is insufficient to explain relations between the inscriptions. Instead, the main text reads 'the gravity of the earth pulls the object in a direction perpendicular to the horizontal surface of the earth. . . . The magnitude of gravity on the moon is different from that on the earth. . . . The weight is different according to the location where it is measured, but the mass is always the same'. The table accompanying the text gives the values of relative gravities on the Earth (1.00), the moon

(0.17), Mars (0.38), and other planets. The main text and the table need to be seen as corresponding to the contents of figure 7.3, and they have to be interpreted (st1, st2), as, for example, articulating the difference between weight and mass, before they would be linked to the inscriptions.

To understand the inscription as it was intended—i.e., to get its point—all three inscriptions and the text need to be connected by the reader. The drawings of the balances and scales, placed on top of the drawings of the earth and the moon (background), have to be perceived as if they were located on the earth (in the case of the drawings on the left) and on the moon (on the drawings on the right) (tl8, tl9). The earth and the moon (background drawings) have to be articulated as if they were pulling objects towards their surface, with different gravities (text, table) (tl2). The blue arrows have to be related to the different gravities and their lengths to the different gravitational forces of the earth and the moon (table) (tl4, tl6). In addition, the sameness of the equal-arm balances and the difference on the spring scales have to be translated into the text; the spring scales have to be linked to weights and the equal-arm balances to masses (tl3).

These relations have to be linked together in the form of a sequential translation. That is, the balances and scales (naturalistic drawings), located on the earth and on the moon (background drawings) where gravity is different (text and table), have arrows (drawings) with different lengths (tl7), representing difference in gravity (text and table). Furthermore, the difference in the weight occurs only in the spring scales (naturalistic drawing) in different locations (background drawing and text), so that weight (measured by the two spring scales) is related to gravity, whereas mass (measured by the two equal-arm balances) is not. Although most readers might not have experienced weighing things on the moon, the background drawings and the arrows must be used to imagine such situation on the moon (tl5, tl9). There is little normal readers' experiences can bring in terms of relevant world experiences to assist in this process.

While reading figure 7.3, readers have to rely on previous experiences with cultural conventions regulating the relationship between inscriptions and the things they depict. For example, the reduced scale (ratio) between the drawings of the balances and scales and the earth and moon must be interpreted as the result of a zooming process (this translation work corresponds to the magnification dimension). The relative size (diameter) of the earth and the moon is different from the real proportion (the moon is one quarter of the size of the earth). In addition, no arrows are perpendicular to the surface of the earth or the moon. These perceptions have to be ignored, as they constitute, from the perspective of someone who already knows them, trivial limits of inscriptions (tl8). Alterna-

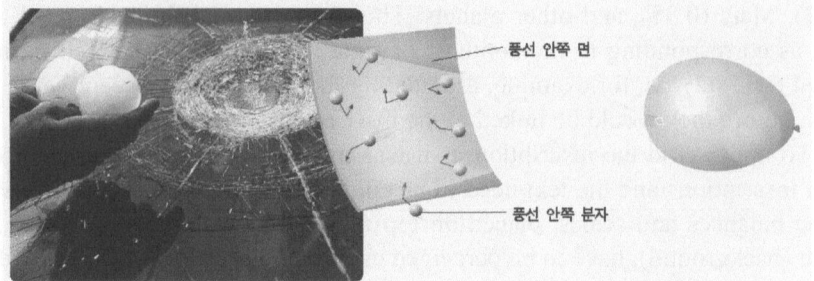

Fig 11. The collisions of molecules. The molecules of gas inside a balloon collide with
the wall of the balloon, in the way the hailstones collided with the windshield of a car,
causing an impacting force.

*Figure 7.5. Example of analogical layered inscriptions. The letters on top read 'inside
wall of the balloon' and at the bottom read 'molecules inside the balloon'. (Reprinted
with permission of the copyright holder. The translation of the caption from Korean is
ours.)*

tively, teachers should be aware of these possibly confusing problems related to
reading inscriptions. Furthermore, the blue arrows are placed only on the equal-
arm balances. The two spring scales have no arrows, which could have otherwise helped readers to see the difference between the two spring scales, as the
lengths of the arrows (gravitational forces) would be different. Nevertheless, if
these background drawings and arrows were not used, readers might be per-
plexed to see that the same spring scale measures the same objects differently.
The background drawings and the arrows are resources that provide context to
the different balances and scales, thus potentially helpful to readers in making
sense of the different contexts of measurement (tl1).

Analogical layer

As their name indicates, analogical layers present a perceptual analogy, intended
to assist reading inscriptions and learning the lessons the inscriptions are designed to teach. In figure 7.5, the photograph of the hailstones and a broken
windshield constitute an analogy for how air molecules hit the inside of a balloon. The stated purpose of the inscription is to allow students to understand the
pressure of a gas as the result of the collisions of air molecules with the wall of a
given vessel. Figure 7.6 shows the work of reading figure 7.5 required for
achieving the intended pedagogical purpose of the inscription. The filled balloon,

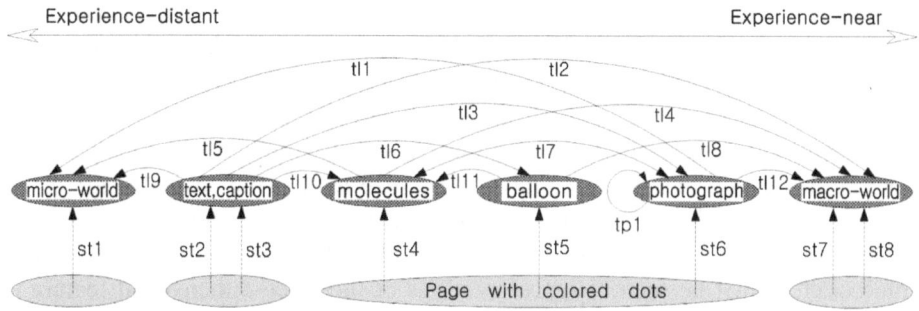

Figure 7.6. Works required for reading the layered inscriptions in Figure 7.5. Each arrow represents the work of structuring ([st], dotted arrows), transposing ([tp], circular arrows), or translating ([tl], curved line arrows).

which is the macroscopic phenomenon (far right in figure 7.6), is to be explained by the collision of invisible molecules on the balloon's wall, represented in the molecular diagram. The microscopic world (far left in figure 7.6) is also to be constructed with the molecular diagram and its analogous photograph.

To learn the intended lesson, initial structuring has to start with the separation of three kinds of inscriptions in figure 7.5: Naturalistic drawing of the balloon (right), molecular diagram (middle), and photograph (left). Pale blue lines and shadows between the diagram and the drawing have to be perceived as a magnification of a small area (white dotted square) of the drawing (the balloon) (tl11 in figure 7.6). Eight small blue circles with shadows and a black arrow attached to each one of them have to be seen as part of the pink plate that is separated from the rectangular photograph and represents the magnified part of the balloon (st4).

At a more fine-grained level, the drawing of the balloon has to be articulated as an air-filled balloon; the two-dimensional oval shape with pink and white color gradients has to be perceived as a balloon, inflated and tied at its opening (st5). This color gradient may assist readers to perceptually structure the drawing to include perspective, thought to assist in making the connection with their experiences in the real world. That is, the white light reflected on the left frontal area of the balloon would account for such color gradient on the surface of the pink three-dimensional balloon (tl8).

Perceptual structuring of the middle diagram reveals that the pink plate and the small blue (meteor-like) circles are 'the inside wall of the balloon' and 'the molecules inside the balloon' respectively (st4). Although the letters 'the inside wall of the balloon' are connected by a black line to the dark-pink triangular plate, the whole pink plate of rectangular shape has to be seen as *a* continuous surface (st4). The curved lines of the pink plate and white-pink gradient have to be seen as depicting a curved concave surface (st4). Similarly, the seven meteor-like circles inside the pink area have to be seen as 'the molecules inside of the balloon', though the corresponding Korean character is connected to the small blue circle outside the pink plate (st4). The blue shadows attached to the small blue circles must be seen as representations of the three-dimensional feature of balls, with tails originated from movement (st4). As such, the tail-like white-blue shadows accompanying every ball should be perceived as depicting the trajectories of the balls (molecules) inside the balloon. That is, the balls have to be seen as moving in the opposite directions of their tails (st4). Similarly, the broken black arrows have to be perceived as the future trajectories of each ball, although in some cases the direction of the past trajectories and the future ones do not parallel each other. Therefore, the balls must be perceived as moving into the wall of the balloon at the points where the black arrows are broken, changing their trajectories towards the direction indicated by the tip of the arrows (st4); this way of perceiving it would lead us to the intended lesson. This perceptual structuring of the middle diagram is consistent with the caption that reads, 'the molecules of gas inside a balloon collide into the wall of the balloon' (st3, tl10).

Perceptual structuring of the photograph is not outwardly self-evident, because the photograph shows a peculiar phenomenon. Most readers, especially those who have not experienced such phenomena, have to resort to the caption, 'hailstones collide with the windshield of a car, causing an impacting force'. With this caption, readers have to articulate or imagine the photograph as a car windshield broken by the falling hailstones (st3, tl3). Within-inscription comparison (tp1) reveals that the sizes of the two big hailstones are about half of what is assumed to be an adult's hands, and they are similar in size and shape to the impact site left on the broken windshield which is apparently produced by a falling hailstone (st6). The blue color of windshield has to be seen as the refracted color of the sky (st6). If the readers have had such an experience, they would be able to notice that the special characteristic of the car windshield glass does not allow it to crack sharply, but produce the white circular shapes when hit by the hailstones (tl12).

The three kinds of inscriptions have to be linked (translating work) to each other, to the text, and to the real (macroscopic and microscopic) world. The pink plate of the diagram has to be related to the drawing of the balloon, as it depicts the enlarged (and flipped over) inside part of the balloon (tl11). Readers have to recognize that this magnifying or zooming is an imaginary process different from the processes in usual optical instruments such as a telescope or a microscope (tl5). We cannot see the molecules of air inside the balloon, and we cannot slice the wall of a balloon in the way it is schematically represented in the diagram. Readers are asked to see the air molecules inside the balloon, from a microscopic viewpoint (tl5). Thus, the relationship between the molecular diagram and the drawing of the balloon is a kind of explanatory layer.

The photograph and the diagram have to be linked through an analogical relationship. This analogical translation requires much work. There are few similarities between the photograph and the diagram (except for the circular shape of both the hailstones and the blue molecules). The words 'in the way', which appear in the caption, have to be used as a clue for translating analogically from the photograph to the diagram, and from the diagram to the photograph. That is, this translation is bi-directional (see the arrow of tl7). The hailstones have to be mapped onto the air molecules, and the (outside) car windshield must be mapped on to the (inside) wall of the balloon (tl7). These mappings are rather explicit or direct in the sense that they can only be done by relating those elements of the two inscriptions with the surface features of two figures. In addition, another kind of mapping, implicit or hidden, has to be made, such as mapping the moments of collisions in each inscription or the mapping of the impacts of those collisions (tl7). This last mapping exercise is the most important aspect of the work done to understand pressure in a microscopic sense. That is, the movement of the air molecules has to be articulated as it results in the molecules' collisions into the inner wall of the balloon and also results in the impact of forces on the balloon's inner wall (tl7).

These translations (mappings) can be affirmed while reading the text. Following the figure, the main text states, 'the molecules inside the balloon move around freely, colliding continuously against each other and against the wall of the balloon. The balloon inflates by the impact force resulting from the collisions of the molecules against the wall of the balloon'. Some part of the main text has to be structured before they are linked to the layered inscriptions (st2). For example, readers must imagine the situation in which two molecules collide against each other, because such collisions between molecules are not depicted in the inscription (tl5). With the main text, it has to be concluded that the impact force of the collisions of the molecules against the wall of the balloon is like the one of the hailstones crashing on a car windshield (tl9, tl1), and that such force inflates a balloon (tl2, tl6).

Translating between layered inscription and real world is a two-fold process in figure 7.5. The real world has to be seen (st1, st7, st8) as having microscopic aspects accessible only by means of mediating tools and instruments. In addition, two kinds of macroscopic world experiences of the readers have to be activated: experiences with an air-filled balloon and with hailstones (st7, st8). The drawing of the balloon has to be related to the experience with the air-filled balloon (tl8), and the photograph of the hailstones has to be related to the experience with hailstones, or it has to be used to foster imagination of the hailstones falling (tl12). The microscopic world has to be constructed by readers in relation to the diagram (st1 and tl5). The construction of the microscopic world (view) could be assisted with the analogous photograph and related to real world experience with hailstones. At the same time, readers have to ignore some other kinds of analogical relationship that may exist between the photograph and the diagram. The mapping might be done between structural features of the two inscriptions, which are not relevant to the scientific concept being presented. For example, the hailstone always falls downward whereas molecules move in all directions. Therefore, these irrelevant mappings have to be distinguished from the relevant concept and then ignored (tl7).

In sum, the drawing of the balloon (the macroscopic phenomena) has to be linked with the diagram (the microscopic world) with the aid of the analogous photograph and the main text. The analogical layer between the photograph and the diagram, and the explanatory layer between the diagram and the drawing have the potential to help readers associate the macroscopic phenomenon with the microscopic viewpoint (tl4 and tl5), which is one of the most important concepts in chemistry.

Data-presenting layer

Data-presenting layers often consist of tables in which several cells are filled with photographs or drawings as examples of phenomena. Figure 7.7 features a table of six major rock-forming minerals. This table is a complimentary resource located at the end of the unit on the topic of 'identification of minerals and their properties'. No more explanation is provided about the contents of this table. Six kinds of minerals are tabulated with their appearances, colors, split types, hardness, and crystal forms. The row of appearances is filled with photographs of each mineral, and the row of crystal forms with diagrams depicting the shape of crystal that constitutes each mineral. Because the textbook authors constructed the tables, these do not have the authentic quality of a photograph in the way we presented them in chapters 4 through 6. The photographs inserted in a table can add more authenticity to the table, providing visual information, such as, for example, color and shape of the minerals, which are important characteristics of

석영	장석	흑운모	각섬석	휘석	감람석

Figure 7.7. Example of data-giving layered inscriptions. The table summarizes the properties of six rock-forming minerals including, from left to right, quartz, feldspar, biotite, hornblende, pyroxene, and olivine. The rows are, from top to bottom, appearance, color, split type, hardness, and crystal form. (Reprinted with permission of the copyright holder.)

the aspect 'appearance'. But in their specificity, the photographs also hinder the identification of these minerals when actual samples differ in some aspect from the tabulated photograph (see chapters 4 and 6).

Reading tables is hardly ever taught in school science and teachers appear to assume that students already are competent in this required reading practice. In figure 7.7, photographs and diagrams can be seen as enlivening the table, which otherwise would be in the form of an aggregation of words and numbers. This layered inscription is intended to teach students to identify six kinds of minerals and compare their properties. The photographs featured are nearer to the experience of the everyday world (right in figure 7.8), and diagrams and properties of words and numbers are more experience distant (left in figure 7.8). But how do these features mediate the work of reading, which we model in figure 7.8?

At a general level, the inscription consists of three tables. The top row has six brown rectangular shapes containing text; the next row presents six rectangular photographs and one pale blue elongated shape with rounded extremities (left

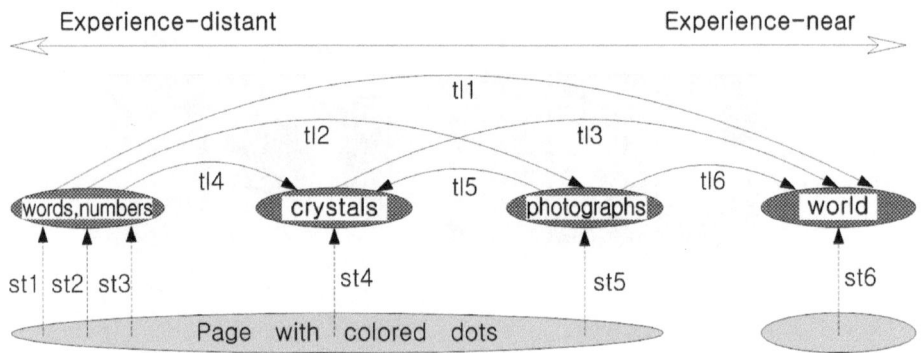

Figure 7.8. Works required for reading the layered inscriptions in Figure 7.7. Each arrow represents the work of structuring ([st], dotted arrows), or translating ([tl], curved line arrows).

side), with textual information inside. Under these rows, a black lined rectangle includes six columns and three rows, filled with letters and numbers. Outside this rectangle, one can see three circles with different colors and shapes, again containing written information. In the bottom row, six diagrams are placed inside a black lined rectangle and one pale violet rectangular shape with rounded extremities is placed outside it, on the left side. White lines divide the two black lined rectangles, creating six columns and three rows in the upper rectangle, and six columns in the bottom rectangle, whereas the two top rows are separated by narrow spaces. Although there are three separate tables, figure 7.7 must be perceived as *one table* with two dimensions of columns and rows: The words in the six brown rectangles on the top row name each column; the words in the leftmost column name each row. That is, reading the cells in one column, one has to assume that something is constant (st1 to st5 in figure 7.8). For example, the column under the first brown rectangle, which consists of a white–blue photograph, the words 'no color', 'no split', '7', and a white–violet diagram (from top to bottom row, respectively) has to be perceived as properties of 'quartz'.

The six photographs in the second row have different background colors. The selection of background color depends on the color of each mineral. If an inscription is appropriately structured into figure and ground, the background highlights the object in the foreground without itself becoming the focus of attention (see chapter 4). For this table to take us to the identification of minerals, the foreground (minerals) has to be distinguished from the background and then has to be compared to other items in the series of photographs (st5). Comparing the six photographs reveals that the table gives independent properties (e.g., appearances) of the six minerals, rather than gradual variations between properties (st5). The six diagrams share the same background color. Perceptual structuring of these six diagrams has to be focused on comparing the six figures (e.g., crystal forms). This comparison reveals that the six diagrams have different (independent) shapes and colors (st4). In addition, each diagram has a white gradient on its foreground, which has to be perceived as the reflection of light on the surface of a three-dimensional crystal (st4). Similar column-to-column comparisons have to be done with the other rows (st1 to st3). In the 'color' row we can read 'no color', 'white', 'black', 'green-brown', 'green-black', and 'yellow-green' from left to right in each column. In the 'split type' row the text in each column reads 'no split', 'two directional', 'one directional', 'two directional', 'two directional', and 'irregular'. It has to be concluded that there are no apparent linearity or regularities in the rows named 'color', 'split type' and 'hardness'.

Identifying the six minerals requires the comparison between rows, through between-inscription-type comparisons. Comparing them with the six diagrams assists the perceptual structuring of the six photographs. The six diagrams have to be seen as reductions or simplifications of the photographs (tl5). Comparing the photograph-diagram pairs in each column reveals that such transformation processes included filtering (in feldspar, one unit of crystal is extracted on the diagram from a number of crystals in the photograph), uniforming (in hornblende, the texture of the surface of the photograph is evened out), and upgrading (in pyroxene, the borders of the crystal are cleared in the diagram). All of these actions constitute work that went into the preparation of inscriptions, and now requires work to move in the reverse direction.

In addition, the translations (comparisons) between photographs and diagrams and the color row have to be done. The color row has to be articulated as presenting the colors of the rock-forming minerals (crystals), instead of the colors of the background (tl2). Although comparing the color of a given figure and the word depicting that color seems to be an automatic process, in some cases, this might generate confusion: does quartz have no color? Is feldspar black? In feldspar, the color of the crystal in the diagram is uniformly black, while the color of the feldspar in the photograph is a mixture of black, yellow, and white. If the term 'no color' means transparency, then the quartz crystal in the photograph or in the diagram should be of the same color as its background. However, the color of the quartz in the photograph is white and violet, whereas the color of the background in the photograph is blue. To solve this puzzle, readers have to understand how the photographs were made much like the scientists in our studies on graphing had to know the process that generated a graph to be able to interpret it. Readers have to assume that the photograph of the quartz was taken, for example, against a background using violet and white light, and then, later, the crystal was moved and placed against a blue background. In the case of the feldspar, the yellow-white color has to be seen as a refraction of light (st6). Finally, systematical comparisons of the relationships between split type or hardness and the shape of the mineral (crystal) are required: Is there regularity between them? Perceptual structuring and translation reveal that there is no apparent dependency between the split type and the shape of the mineral, no possible mathematical or logical relationship between the numbers of hardness and the other properties (tl2, tl4, tl5).

In sum, the table in figure 7.7 has to be structured with its columns and rows and comparisons have to be made between photographs, diagrams, words, and numbers, in and between columns and rows. Each row and column has to be translated into the reader's real world experiences on the properties and observations of real rock-forming minerals (tl1, tl3, and tl6).

Explanatory layer

The most frequent type of layered inscriptions in both Korean and North American textbooks is based on the explanatory layer, here exemplified in figure 7.9. The photograph on the left presents a two-headed train, with two arrows of different colors just above it, with letters and numbers above each arrow. The figure at right presents several parallel arrows, again with letters and numbers above each vector, and the equation '$F = F_1 + F_2$'. The intended lesson to be learned in reading the layered inscription in figure 7.9 is that forces in the same direction add in the way vectors (arrows) add. Figure 7.10 shows the work of reading figure 7.9 required for arriving at the targeted lesson. The different layers are designed to provide readers with resources to bridge the gap between their experienced world (far right in figure 7.10), more easily associated with the photograph of the train, and the mathematical equation (next far left in figure 7.10), the inscription farthest removed from lived experience in the world.

Initial perceptual work is required to separate different types of inscriptions: Photograph, arrows in the photograph, diagram of arrows, and mathematical equation in the diagram. Then the work of structuring, transposing, and translating have to be done. The (static) photograph must be perceived as depicting a train with two head engines moving from left to right (st8 [figure 7.10]). Although in some cases the train moves backward, such cases have to be disregarded (tl11, st9).

▲그림21 같은 방향으로 작용하는 두 힘의 합성   합력 F는 F과 F를 합한 것과 같다

Fig 21 **Addition of forces in the same direction.** The resultant force (F) equals the sum of F_1 and F_2.

Figure 7.9. Example of explanatory layered inscriptions. (Reprinted with permission of the copyright holder. The English translation of the caption is ours.)

In addition, the two head engines have to be articulated as together exerting forces in the same direction to pull the whole train (st8). There are special elements (other inscriptions) other than the photographic image that have been added after the photograph was developed: The two arrows with letters and numbers above them. The two arrows have to be perceived as being parallel to the train, pointing in the same direction (left to right) (st6, st7). The short blue arrow has to be structured as corresponding to both the second (left) head engine and to the letter and number combination F_1 (tl10, st6). The long green arrow corresponds to both the first (right) head engine and F_2 (tl10, st7). However, the caption does not state directly what these annotations mean (st9). Instead, the main text states, 'Thus, when there are forces in the same direction, the resultant force (F) is equal to the sum of one force (F_1) and another force (F_2). The direction of the resultant force is the same as the direction of the two forces that are added up'. With this text, the letter-number combinations can be translated into forces, and each arrow can be linked to the force of each head engine (tl3). In addition, two arrows have to be articulated as representing the directions of each force (e.g., head car), and the different lengths of the two arrows have to be inferred as representing the different magnitudes of each force, although the latter is not mentioned anywhere (tl10, st6, st7).

The diagram on the right, containing several parallel arrows requires complex structuring and transposing. The background of the diagram has to be seen as a graph paper (st3 to st5). The lines and arrows of the diagram are drawn onto

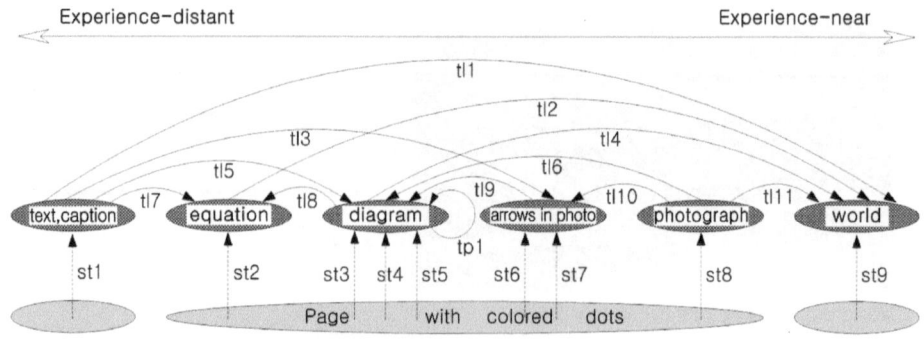

Figure 7.10. Work required for reading the layered inscriptions in Figure 7.9. Each arrow represents the work of structuring ([st], dotted arrows), transposing ([tp], circular arrows), or translating ([tl], curved line arrows).

the lines of the graph paper, providing us with a criterion for comparing the different arrows: All arrows have different lengths but the same direction (st3 to st5). The black vertical line with three 'zeros' vertically aligned beside it must be perceived as the ordinate of the graph (st3 to st5). Also, the tails of the arrows begin at this line aligned like the starting blocks of sprinters. In addition, arrows are associated with numbered letters except for the red arrow on the bottom, associated with the equation '$F = F_1 + F_2$' below it. In the third row of the diagram, the green arrow starts from the tip of the blue arrow. There is another vertical dotted line that is parallel to the black vertical line, from the tip of the top blue arrow to the middle of the bottom red arrow (st3 to st5). Comparing color and lengths of the arrows reveals that the two blue arrows in the first and the third rows of the diagram represent the same entities but at different locations (tp1). Thus, the first arrow is actually the third (blue) arrow, just placed in a different location on the diagram. Similarly, the two green arrows have to be articulated as being the same entity (tp1). Comparing the length of the red arrow and the two arrows in the third row reveals that the red arrow has the same direction of the blue and green arrows, and the length of red arrow is equal to the sum of the lengths of the blue and the green arrows, which are summed graphically in the third row of the diagram (tp1).

Finally, the equation, here the endpoint and most generalizable aspect of a long cascade of inscriptions, has to be structured as it makes equivalent 'F' with '$F_1 + F_2$' (st2). This equation has to be articulated as corresponding to the red arrow, the sum of the blue and green arrows (tl8). However, the graph paper is an arbitrary mathematical space. Therefore, translating is required from all the arrows in the diagram (and the equation) to the photograph (closer to the real world) and caption or main text, for figure 7.9 to lead us to the targeted lesson.

The blue and green arrows in the diagram and the photograph have to be articulated as referring to the same entity, although the colors of the letters and the (green) arrows are different (tl5). The arrows in the photograph have to be perceived as being the same arrows as those in the diagram, therefore with the same length and direction, and most importantly, representing the driving forces of the two head cars (tl6, tl9, and tl10). The color and relative lengths of the arrows are linking cues that can be seen as relating photograph and diagram. However, the directions in which the train moves and in which the arrows point are not the same. The train is located at an angle with the bottom of the photo. The classical conventions of painting and photography associate the distinctions high versus low with far versus near. Thus, the train might be seen as moving from left to right and also from far to near (in relation to a point located outside the photograph, in this case, the reader). Readers have to ignore this difference of direction between the photograph and the diagram (tl10). Nevertheless, the photograph gives the readers an everyday (authentic) example of forces with the same direction. The diagram depicting the resultant force has to be translated to the readers' everyday experiences via the photograph with the same kind of arrows (tl4). In this way, readers have to conclude that forces with the same direction (text and caption) can be added not only in a graphic form (diagram or photograph) (tl5) or mathematical form (equation) (tl7), but also in a situation from everyday experience (tl1). In addition, students are to learn that the arrows— depicting forces arranged in a horizontal line along the train in the photograph— can be re-arranged on the diagram vertically, in parallel, and can be back-arranged into one line (st3 to st5).

Graphicacy and pedagogy of layered inscriptions

In the past, scholars of many disciplines—including philosophy, cognitive science, and artificial intelligence—thought about the relationship between language and world in terms of two different arbitrarily related domains. Recent work among scientists showed, however, that scientific research translates world into language by translating and structuring matter along a chain of inscriptions separated by ontological gaps (see chapter 4, especially figure 4.2). In reading a scientific text with inscriptions, readers are asked to take the inverse journey, moving from language and experience-distant inscriptions to the world, to get their point. Layered inscriptions provide reading resources that break the larger gap into many smaller ones. Our semantic model captures the ordering of the inscriptions and the different kinds and amounts of work required for differentiating and relating the layers to one another and to the lived experience of the reader. In this chapter, we describe the work required to learn from inscriptions, in particular, composite inscriptions in which multiple types of inscriptions are layered; that is, we describe the special dimensions of *interpretive* graphicacy required for reading inscriptions in which their authors combine conceptually different layers. Except for simple inscriptions, layering provides the basis from which new meanings can emerge in and between inscriptions. How do different layers mediate reading and thereby assist rather than hinder students in bridging generalizable cultural knowledge to personal experiences?

Our analyses provide examples of three kinds of work required for the intended learning outcomes: Graphicacy involves structuring and transposing within a layer and translating between layers and between inscription and the lived experience of the reader. The additional layers of inscriptions break the overall gap between the most experience-distant inscription and lived experience into smaller gaps. However, the gaps do not disappear and continue to require work. The proliferation of arrows in our model shows that the additional layers require additional structuring, transposing, and translating work. We should therefore not be surprised when students find difficult to read and interpret these layered inscriptions, even though they were designed to facilitate learning.

The present analyses show how many different interpretive actions (structuring, transposing, translating) have to be done for the pedagogy of the inscription to work. However, signs or symbols do not have one specific sense or meaning—the very idea of a sign (symbol, inscription, word) standing for something else *implies* multiple senses.[5] In the light of our findings that brought out a proliferation of relations between different inscriptions that a student has to interpret in the one rather than any way to get the point, layered inscriptions may actually make reading more difficult. This is also true when the inscriptions bear iconic relations to their referents (e.g., a photograph of a train and a real train), which are made by convention and therefore need to be learned. Students therefore need to develop knowledgeable graphicacy to interpret additional visual inscription. Whether and how this is the case requires empirical studies investigating students in the process of interpreting layered inscriptions in textbooks. Our study provides a framework to direct the attention of investigators doing either qualitative or quantitative research.

The interpretation of inscriptions always requires structuring and translating work; these processes are an integral aspect of knowledgeable graphicacy. Transposing work is required only when there are two or more inscriptions of the same kind to be compared with one other (e.g., figures 7.3 and 7.9). With layered inscriptions, not as a single inscription, we focus on the translation work. This work depends on the functional relation between inscriptions. Here, the four examples analyzed exhibit differences. When a background layer is used, the work of translating takes the form of zooming (magnifying) and locating (or imagining) other inscriptions in a specific context; with an analogical layer, the corresponding work consists of active construction of analogies; comparing and filtering constitute the brunt of the work when the additional layer presents data; and some pictorial form of everyday example or experience is used in an explanatory layer. Thus, different kinds of layered inscriptions require different kinds and amounts of work. In addition, the translating works add new meaning (for example, provide context for the inscription with the background layer) that would be impossible if there were no layer.

Layered inscriptions have a pedagogical potential for assisting students in learning about a new topic. In particular, the different layers decrease the length of the steps that have to be taken from the most experience-distant inscription to the lived world of the reader. They give additional resources to readers such as, for example, the context of an inscription, analogical material, everyday example, or scientific phenomena. However, they also require more steps to be taken, which may complicate learning. Our analysis showed the tremendous amount of work required unpacking an inscription. Our model of graphicacy with respect to layered inscriptions articulates the different types and amounts of work involved in reading such inscriptions and arriving at some new understanding.

Our analyses allow us to ask new questions. For example, what are the types of inscription that facilitate versus impede reading? Would alternative layers in our examples have provided advantages? For example, in figure 7.1, is the work more or less if the photograph of syringe and piston were used rather than the drawing of beaker and piston? A photograph requires structuring and translating similar to the drawing, but it is closer to real world experiences and thereby might allow students to bridge the first gap more easily. New studies, in which students are asked to read layered inscriptions, are required for determining the difficulties associated with each type of work so that the different arrows can be associated with difficulty levels. Our model lends itself to preparing experimental studies in which the different types of work are systematically varied.

Our model allows textbook authors to identify the different types of work required for the reading to unpack the lesson embodied in an inscription. Authors should consider not only all of this work in trying to make layered inscriptions, but also they should consider the related social conventions or prerequisites (experiences) that are required. Authors need to be aware that not a single aspect of an inscription can be taken as self-evident but that it requires work to unpack the lesson it contains. Sufficient resources should be provided with each layered inscriptions, such as guiding (linking) lines, arrows, and specific colors, and providing detailed explanations in the caption and in the main text. Some aspects of this work may be more difficult than others.

Our model also guides teachers in identifying the various kinds of work involved in reading layered inscriptions. The analysis of the work required in interpreting specific layered inscriptions can help teachers arranging the different types of work in their lessons, helping students' learning. For example, the teacher can list three kinds of works to be done, and guide students perform those works, focusing on some important or difficult ones. Alternatively, the sequence of types of work can be used as a mean to facilitate students' discourse around layered inscriptions, fostering active participation in reading them together. Or the teacher could provide students with real world experiences (for example, real rock-forming minerals for figure 7.7) that are related and required in reading inscriptions.

8   Semiotics of chemical inscriptions

Thinking about macroscopic phenomena in terms of models based on the idea of microscopic particles (i.e., the particulate theory of matter) is one of the important goals for student learning in chemistry. However, previous research suggests that students do not easily understand phenomena from a particle perspective, although such a perspective has many concrete aspects that ought to assist learners of chemistry. More than the textbooks of other countries, Korean chemistry texts include colorful inscriptions. How, we might ask, do such inscriptions help learners of chemistry? The purpose of this investigation is to investigate the function and structure of chemical inscriptions in middle school science textbooks by drawing on a semiotic framework. We develop the concept of *chemisemiotics* to unveil the work of reading required to understand chemical inscriptions in the way their authors intended them to be understood. We begin with the assumption that different kinds and functions (structure) of inscriptions constitute different signs in the learning process. In contrast to the standard approach, which attributes learning difficulties to students' mental deficiencies, we show that difficulties in understanding chemistry may result from the different processes of semiosis (interpretation and meaning-making) between inscriptions depicting macroscopic and models based on microscopic particles. This study therefore embodies a form of critical graphicacy, which we use to identify the work required in reading for uncovering the politics of representation in chemistry. In particular, we take up representations related to a central theory in chemistry, the particulate theory of matter.

Contradictions in and of inscriptions in chemistry

Science textbooks contain many different kinds of inscriptions that constitute signs or sign complexes in respect to which we want students to develop critical graphicacy. In chemistry, two kinds of inscriptions are typical: those denoting phenomena that can be observed in the surrounding world and those denoting

models based on the idea of invisible microscopic particles. The unacknowledged problem, from our perspective, lies in the fact that the microscopic particles are presented to students as phenomenal facts rather than as models. If this
is the case, it would not be surprising that students experience difficulties. The
chemisemiotics we develop here, itself a form of critical graphicacy, allows us
to uncover these contradictions internal to the discipline of chemistry and
chemical education.

The conflation of the ideas of phenomenal microscopic particles and microscopic particles as parts of a model is evident in the following example. To teach
the differences between solids, liquids, and gases, one Korean textbook uses
photographs of people in different situations and drawings of spherical objects
(figure 8.1). We may ask, 'What kind of work is required in reading and understanding these inscriptions and thereby learning from them?' An analysis of the
compound inscription in this figure might begin as follows. The inscriptions in
figure 8.1, as part of a question, provide the reader with two different representations to be linked by an analogy. The molecular model on the left consists of
three beakers with red balls distributed in three different ways. The photographs
on the right contain different situations where people are standing or walking
during an event at the Olympic games. The purpose of this question is to explain
the different molecular arrangements in the three states of matter.

A response to our question appears to be that students simply need to see
the analogy that exists between the left and right images: students will use one
familiar inscription (say, photographs) to understand another, unfamiliar (molecular model). Considering the fact that people usually see the left first, the molecular model is assumed to be more important, serves as the source structure to
be imposed on the photographs, the target structure of the analogy. In this case,
it might actually be better to exchange the left and right sides, because the right
photographs seem to be more familiar to the students consistent with the princi-
ple that learning goes from the familiar to the unfamiliar. However, before
jumping to quick conclusions and reversing the two sides, let us consider the
relative importance and familiarity of entities depicted.

One can say that the inscriptions on the left side already contain an analogy.
The molecular model draws on a particulate *theory* of matter. The model is not
from some kind of logical, pervasive perspective; in fact, from a physicist's perspective, the balls represent an entirely false picture of matter. What do the balls
represent? Balls are things that exist in the world directly accessible to our experience. It only can be said that the model (diagram) uses the macroscopic phenomena or things (balls) to speculate about what is happening at some microscopic level to which we do not have access. We assume that at the microscopic

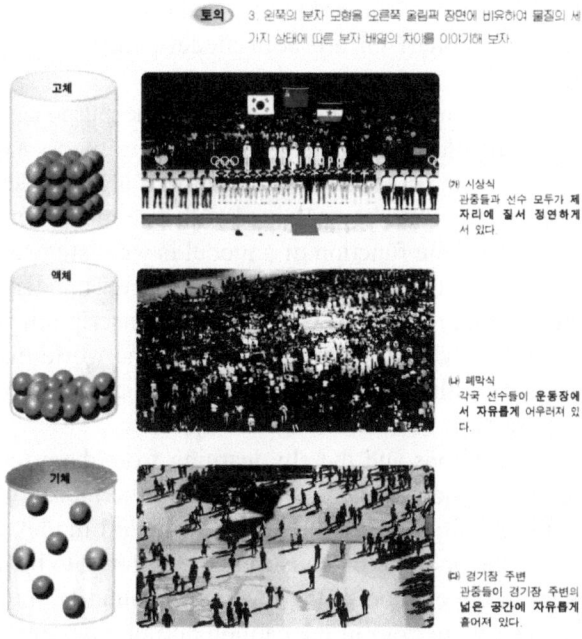

Figure 8.1. Page from a Korean textbook. These inscriptions are part of the question: 'Discussion 3. Drawing on an analogy between the left molecular model and the right scenes during the Olympics, let's talk about the different arrangement of molecules in three states of matter'. The words on the left side are 'solid', 'liquid', and 'gas' from the top respectively; in the right side it reads: '(a) A ceremony of awarding prizes: The audience and the athletes are all in their place standing orderly. *(b) A closing ceremony: Athletes from all over the country are standing mixed* freely on the sports field, *and (c) The outside of the stadium: The audience is scattered* freely in the open area outside of the stadium'. *(The layout of the translated textbook page follows the original. Reprinted with the permission of the copyright holder.)*

level, molecules function similarly to the way balls function at the macroscopic world. Thinking in this way is already analogical in nature; there is no distinction between microscopic and macroscopic otherwise.

On the other hand, many people think the microscopic world is a miniature version of things (balls) in the macroscopic world, rather than those things being part of *a model*; the macroscopic world is presented only in the photographs. With analogical reasoning, we can clarify and describe what the molecular model is depicting. However, one has to ask, 'Can the right-hand photograph be an analogy to the left model? Or are both model and photograph analogies of something else?'

A model is a tool for explanation, serving some purpose. For example, a ball-like model is suitable to explain the phenomena of crystallization. For a physicist, an atom is a set of equations or a set of functions describing natural phenomena. Certain things can be explained with one model, but others cannot. However, in many cases, the function of a model is what students don't learn. In figure 8.1, this point is not clearly articulated anywhere: the distinction between model and reality is not articulated. Worse, the left inscription is not the model of the right photograph, which depicts an aspect of the world that we know and can experience. As a result, students may confuse what is reality and what they should perceive as reality.

Reading the inscriptions and thereby learning from them requires the interpretation of each inscription as well as the text; furthermore, readers must link them. Together, the text (question), the letters in the left model, the captions for the photograph, and the photograph constitute the material to be understood. The emphasized letters in the caption of the top-right photograph highlight what the reader should see as salient both in the caption and photograph. The function of text (words) therefore has to be analyzed in relation to the inscriptions deployed on the same textbook page.

In Korean science textbooks, there are large numbers of inscriptions such as this one. Bright inscriptions, as we have seen in previous chapters, are often intended to motivate students to learn, since every colorful inscription has the potential to call readers' attention. Yet, systematic research on these inscriptions has rarely been pursued in Korea or elsewhere, though some types of inscriptions in science textbooks were studied: graphs (chapter 2) and photographs (chapter 4). In those chapters we analyzed high school textbooks in North America and Brazil, and mainly focused on biology units. We now need to better understand the function of inscriptions in science textbooks of different cultures and different subject matters. Our focus here lies on chemical inscriptions; the first inscriptions that students encounter, which depict the particulate nature of matter, could give them an impression that would form a basis for understanding chemical concepts from then on.

Here we explore the potential of a semiotic framework to study the relations between the phenomenal world and the microworld models depicted by inscriptions in Korean textbooks. Our purpose is to articulate the function and structure of chemical inscriptions in middle school science textbooks. We analyze the drawings of chemical phenomena depicting the particulate nature of matter. As part of our analysis, we develop the concept of *chemisemiotics* to unveil the work of reading required to understand chemical inscriptions in the way they are intended by textbook authors or by the school curriculum. We begin by taking a look at inscriptions in chemistry and the particulate theory of matter, which is part of the central dogma in the discipline.

Inscriptions in chemistry

Chemistry is about substances and their interactions, many of which can be perceived with bare eyes; but its explanations are in terms of entities (atoms, molecules) that would be too small even if they existed in phenomenally accessible form. Consequently, chemical inscriptions can be classified as macroscopic (e.g., experiments and experience), (sub-) microscopic (e.g., molecules and atoms), and symbolic inscriptions (e.g., chemical formulae or equations). The latter form of inscription mediates between the former as 'Mg', for example, may denote either an atom of magnesium (microscopic model) or the substance magnesium (a silvery metal). The three levels of inscriptions of matter and the relationship between them are crucial in learning chemical concepts. However, there is a debate whether the existence of three kinds of inscriptions (multiple representations) actually help in learning chemistry. Many researchers argue that students find it difficult to transfer from one type of inscription to another and find that students focus on the surface features of inscriptions rather than on the underlying concepts.[1] Generally, neither textbooks nor lectures include descriptions of how such transfer or shift should occur. Our questions are, therefore, 'What is the nature of the work of translating (transferring) between two inscriptions?' and 'What is actually needed in translating, and therefore what helps or hampers the translating process?' For example, in figure 8.1, readers might resort to their prior experience of being part of an audience or playing with balls to interpret the photographs and the molecular model. They would have to draw on social and cultural conventions of drawing such as color gradient to perceive the red circle as a ball, and would have to assume that the two-dimensional inscription depicts three-dimensional space.

Particulate theory of matter

Teaching and learning about the particulate nature of matter has been investigated according to two perspectives: (a) identifying students' misconceptions and their possible causes; and (b) developing teaching and learning methods using drawings, analogies, activities and demonstrations, and computers utilizing multimodal figures. These different approaches resulted in specific implications for the practice of chemistry teaching. However, there is mounting evidence that many students 'lack understandings' of the particulate theory of matter, and that those non-scientific understandings persist even after interventions.[2] Our efforts to work toward critical graphicacy is in part motivated to develop the grounds for critiquing such studies that end up with deficit descriptions of students. Because practice makes practice, it is more likely that students' difficulties arise from the lack of opportunities to participate in the practices that constitute *critical* graphicacy. The supposed lack of understanding is not the students' fault, a result of their cognitive deficits, but a result of science education pedagogy.

Students' difficulties in learning the particle view of chemistry have been attributed to the fact that the particulate or submicroscopic nature of matter contradicts students' intuitive and everyday views of matter as something continuous. There is no need for thinking with a microscopic perspective in everyday life, such as, for example, when they are nailing wood. The discontinuity of matter, along with the concept of vacuum, is the most fundamental assumption in chemistry, which is hard to accept. Students are said to think that there is air (no empty space) between particles, and do not consider the intrinsic motions of particles and interactions (forces) between particles.[3] Some students are said to believe that molecules of the same substance have different size or shape in different phases, and that atoms are alive. Sometimes students attribute macroscopic properties (e.g., color) to single atoms. Students' personal experience, peer culture, language, and instructional materials have been suggested as the sources of such thoughts, beliefs and attributions.

The invisibility of the particles forced the use of a model (symbol) to help people visualize the microworld. However, this model itself gives rise to a series of problems. The model does not solve the problem of the ontological and epistemological nature of the chemical concept, and it has no fixed definition or usage. For example, it is very hard for many students to attribute material properties (e.g., mass and volume) to an invisible entity such as a particle. The model itself does not explain this prerequisite concept. Furthermore, many kinds of models and symbols give rise to further translation processes between them. These issues are apparent in the different inscriptions that we excerpted from a cartoon of a story on the water cycle in one Korean textbook (figure 8.2). Can you distinguish which one of the drawings represents the gas state or the liquid state? The drawings in the middle and on the right side can both be though of as depicting the liquid state of water; the one on the right having a higher temperature than the one in the middle. According to physicist Erwin Schrödinger, the average distance between gas molecules is about ten times their diameter, at

 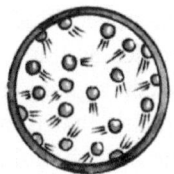 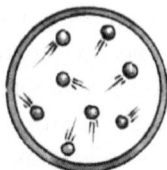

Figure 8.2. Molecular drawings depicting water molecules in solid, liquid, and gas states that appear in Korean middle-school textbooks (Reprinted with permission of the copyright holder.)

least when at normal air pressure and room temperature. However, few drawings in seventh-grade science textbooks in Korea (if any) use this proportion when depicting the gas state at the molecular level. In addition, in these drawings there are no explanations about what the circles and short lines near each circle denote, and there is no explanation of the differences between the colors of the boundaries. That is, the nature or function of this model is not articulated in the textbook. Are these models to be used as *explanations* or *descriptions*? Or should the reader study this model as an object to be memorized rather than to be understood?

There are differences between school science and students' everyday understandings of models. In science, models stand for things that cannot be observed, known or unknown. The function of a model is to assist in conceptualizing the target. Therefore, models can be tested, changed, and treated as more or less adequate. However, teachers usually present a model to be learned as a static fact. Students tend to relate to the model as concrete object and think of the particle model in the same way as of concrete replicas (images) of everyday entities.

Towards a chemisemiotics

Specialized semiotics (studies of signs) have been proposed in different domains and for different organisms, including biosemiotics (communication and signification in living systems), zoosemiotics (communication of animals with their environment), and phytosemiotics (communication of plants with their environments). Here we present a semiotics of chemical inscriptions, to which we refer as *chemisemiotics*, in recognition of similar efforts in other domains. We begin the construction of our chemisemiotic inscription with the Peircean concept of the *sign* (see chapter 1). A sign is something (e.g., molecular diagram) that stands for something else, the *referent* (e.g., a gas molecule in general). The relation between sign and referent is never direct, because a gas molecule in general covers all specific gas molecules and any possible gas molecules not concretely realized. The relation between sign and referent, however, is mediated by *interpretants* (e.g., scientific language that translate or elaborate the sign). For example, if our sign is 'a molecule of carbon dioxide', a drawing containing two (blue) circles standing for oxygen and one (black) circle standing for carbon constitutes an interpretant; other interpretants are the chemical formula CO_2 or the structural representation $O=C=O$ (where each pair of lines corresponds to two pairs of electrons shared between neighboring atoms).

Complementing the presentation of semiotics in much of the literature, we assume in this investigation that there is a reader (or a student) who interprets a given sign in a textbook, and that possible interpretants can be identified that are maximally required in a process of interpreting the sign in a way intended by the author or by the curricular purpose. For example, the inscriptions in figure 8.1 serve as signs to be interpreted using analogical reasoning. We call any unit of the interpretive process (semiosis) *reading work*.

Generally, semiotics is concerned with the codes underlying communication. Here we are concerned with discovering the code of chemical inscriptions. We expect that the code that allows scientists (or science teachers) to uncover sense from textbooks is not available to students, who bring from their daily life a range of interpretants to the reading of chemical inscriptions that differ from those used by more experienced insiders in a community of users. As researchers and science educators, we disassociate ourselves from particular conventions and assumptions that are needed in reading a given inscription, which allows us to identify possible interpretants rather than *the* interpretants characteristic of the discipline. This provides us access to a maximum amount of reading work required in recovering a textbook author's intent.

We extract textbooks elements (inscriptions) to construct semiotic models of reading chemical inscriptions (figure 8.3). Two kinds of inscription are identified: inscriptions that represents elements of the world given to us in everyday experience, for example, a photograph of natural phenomena or an experiment; and inscriptions intended to represent things inaccessible to the naked eye or hypothetical entities, for example, molecules. We cannot say, though, that the second type of inscription denotes the microscopic world, because we do not have access to a microscopic world at the level required to see molecules. However, as we analyze the chemical inscriptions that aim to help the acquisition of the particulate theory of matter, we can say an inscription represents *particles* rather than macroscopic phenomena. Another element is the main text in which chemical inscriptions are embedded, which may include questions, descriptions of ac-

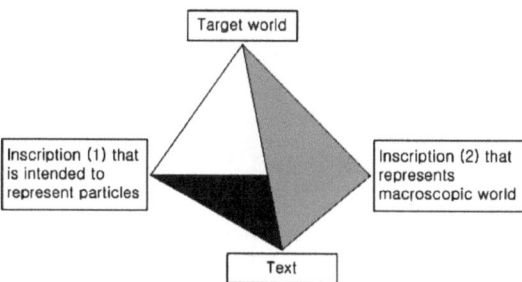

Figure 8.3. A semiotic model for reading chemical inscriptions. Each of the four resulting faces of the triangular pyramid constitutes a triadic semiotic relation.

tivities, captions, and letters (figure 8.3). These three elements (two inscriptions and text) constitute the world of the text (book).

The world constitutes a continuum, including textbooks and ink used to form letters and images. Inscriptions then constitute one aspect of the continuum that are used to refer to or denote other aspects of the continuum. To interpret a sign then means defining that portion of the continuum that serves as a vehicle in its relationship with another portion of the continuum. Therefore, we have target portion of continuum (world), the target world in figure 8.3, as a constituent of our semiotic model. Different inscriptions and text (portions of continuum serving as sign vehicles) pertaining to the same phenomenon (referred to portion of continuum) are put in a relation that forms a triangular pyramid, where the two inscriptions and text are in one plane (the world of the text), and the signified portion of the continuum in another (the world outside the text). The bottom triangle of the pyramid is analogous to the quasi world that text (and inscriptions) makes, and the target world refers to the circumstantial world. Therefore, reading means linking the textbook material (quasi world) to the circumstantial world.

With the pyramid, each of the four faces formed can be read as a triadic relation. For example, the black, bottom plane (figure 8.3) can be structurally analyzed independent of its relation to the (outside) world. The text motivates the inscriptions and the inscriptions validate the content of the text (see chapter 4 on the relations between inscription, caption, and main text). For example, the text in figure 8.1 is in the form of a question; it invites (motivates) the reader to use two inscriptions in solving the interpretive problem. The inscription on the left side provides an image that the text describes as 'molecular model'; the inscrip-

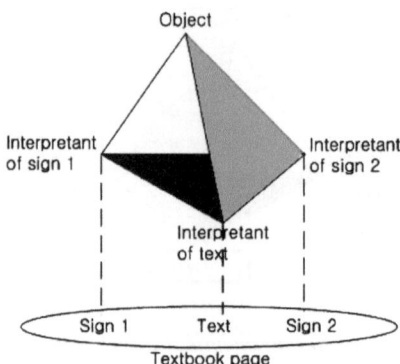

Figure 8.4. A fuller semiotic model including the textbook page, which has to be structured first into signs that then can be interpreted in a second step.

tion on the right side consists of a photograph that shows what the text describes as 'scenes during the Olympics'. The inscription on the left in figure 8.1 is the inscription that is intended to represent the particles of gases, solids, and liquids (figure 8.3), that is, it represents the particulate nature of matter. The other inscription in figure 8.1 represents the macroscopic world likely to be known to (most) students (figure 8.3). The text and inscriptions are the material signs of the textbook page, all referring to the particulate nature of matter in general. No thinkable primary or interpretant signs can directly refer to the particulate nature in general, for each would only constitute one concrete realization of the referent, whereas the referent is general, containing all of its actual and possible concrete realizations.

The text and two inscriptions in figure 8.3 are, actually, the interpretants of those signs that appear in the textbook (figure 8.4). The lower oval in figure 8.4 refers to the material basis of all signs (inscriptions and text) on the textbook page. To unravel the relationships between four elements of our semiotic model, we should think of the interpretant of the three signs (text, inscriptions 1 and 2 in figure 8.3). The relationships between signs are not inherent in signs themselves. We think of the interpretants of signs from a researcher's or textbook author's point of view. The dotted lines correspond to the *structuring* process of analyzing the internal structure of an inscription or text (see the previous chapter on *structuring* work). The black line corresponds to the translation process between two worlds. Reading textbooks (text and inscription) requires both of these forms of work.

The text provides an explanation or a description of the target world. The relationship between inscription 1 and target world is analogical or hypothetical. Readers cannot access any would be particles, the microscopic world, but we have only the theory of particulate nature of matter. Because the microscopic world is invisible, it requires representation by means of things available in the macroscopic world, such as the balls or spheres used in figure 8.1. That is, inscription 1 can be a phenomenological analogy of sensory experiences at the macroscopic level. The relationship between inscription 2 and target world is iconic in the sense that, for example, the diagrams in figure 8.2 show similarity (resemblance, likeness) with the world directly accessible to the (student) reader. (It should not be forgotten that the iconic relationship between a photograph or drawing and some event is something learned.) The relationship between inscription 1 and inscription 2 is not fixed; it varies from one case to another case. For example, in figure 8.1, the relationship between two inscriptions constitutes an analogy; the relationship between figures 8.5 and 8.6 is one between foreground and background.

The target's representation draws on social conventions. In addition, the target contains inner contradictions: the world can be interpreted both in a macroscopic way (as with the inscription 2, photograph) and in a microscopic way (as with the inscription 1, the molecular model). These relationships between microscopic and macroscopic worlds are dialectic in the sense that those two worlds presuppose one another. In a relation between text and inscriptions we can find other contradictions, in which some features are inconsistent across the text and figure. For example, whereas the main text may articulate the *movement* of molecule, an associated inscription cannot show the movement because the figure on a textbook page is always a *static* image.

As implied in our semiotic model, reading chemical inscriptions involves interpretive works both *inside* an inscription (or text), that is, structuring, and *between* inscriptions and text, that is, translating (or transposing). All three elements on the textbook page should be co-organized to get their point. Readers have to make a variety of assumptions and come with specific presuppositions (conventions) in reading signs. The surface feature of one sign may restrict the reading processes; however, the other signs could be used to bridge the gap between one sign and the target world, or used to resolve potential contradictions within and between signs. In the following four sections, each presenting the analysis of one concrete case, we show what kinds of work are required in reading chemical inscriptions, and how our semiotic models are applied to find the chemical codes shared within the community of chemists.

Molecules in layered space

Reading an inscription requires co-organization within it and across other inscriptions and texts. To make sense of an unfamiliar sign, we have to use all the other signs, the co-text of a text. For example, oral or written language is required to make interpretations of chemical inscriptions. As we show in chapter 7, each sign has to be organized or structured, at least in part, before it can be linked to another. In the process of structuring an inscription and translating among other signs, some contradictions will emerge or be resolved. That is, the three elements on the bottom plane in our semiotic model (figure 8.3) help or interfere with each other in being interpreted by the reader. In the following, we articulate these issues in the case of an explanatory and layered inscription that represents a molecular view (figure 8.6).

In this example, the inscription ('Figure 5-12') is placed within the main text and is followed by a question that articulates the figure as a resource for solving a problem, 'How does the rubber balloon placed over the mouth of the bottle change when the internal temperature of the bottle is lower or higher than the external temperature?' The main text describes the situation of a previous experimental activity (figure 8.5)—presented on the preceding page—and provides an explanation of a macroscopic phenomenon (the result of experiment observing the change of a balloon attached to a plastic bottle when the bottom of the plastic bottle is dipped into hot water) using a microscopic perspective. The caption to the inscription provides a similar explanation of the phenomena (figure 8.6). However, because the caption uses different words such as 'temperature' that cannot be found in the main text, this figure can be classified as having explanatory function (see chapter 4). The question below the inscription, which can be thought of as a different but akin genre to the main text, asks about the same phenomenon, but uses different terms or concepts: the difference of temperature (not the change of temperature, as in the caption) between the inside and outside of the bottle. At a global level, all resources (e.g., text, figure, caption, and question) are based on the same phenomena, but with different modalities or signs. Readers are asked to construct their own interpretation of the phenomena using all of these resources together. The instructional intent of the experiment (known to the experienced teacher) and the subsequent text and inscription is for students to understand the relationship between temperature and volume of gas in a particle view.

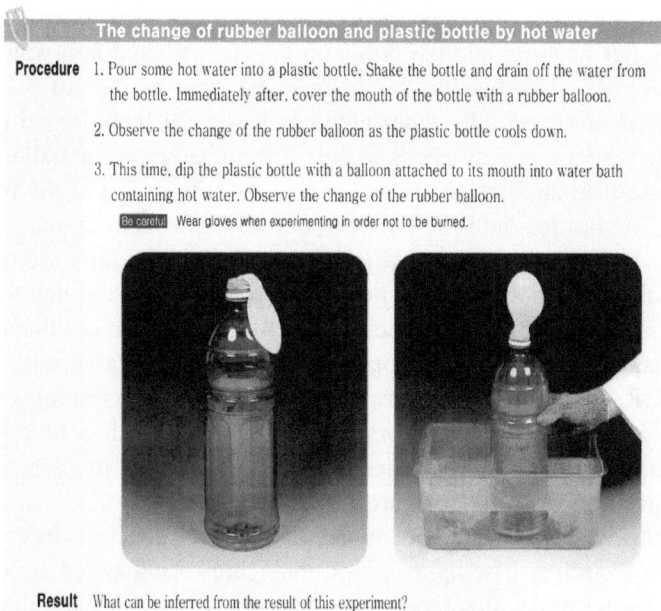

The change of rubber balloon and plastic bottle by hot water

Procedure 1. Pour some hot water into a plastic bottle. Shake the bottle and drain off the water from the bottle. Immediately after, cover the mouth of the bottle with a rubber balloon.

2. Observe the change of the rubber balloon as the plastic bottle cools down.

3. This time, dip the plastic bottle with a balloon attached to its mouth into water bath containing hot water. Observe the change of the rubber balloon.

Be careful Wear gloves when experimenting in order not to be burned.

Result What can be inferred from the result of this experiment?

Figure 8.5. Experiment pertaining to the relationship between temperature and volume of air. It provides students with an experience in the (macroscopic) world accessible to their experience, on which they can draw in making sense of the particulate theory of matter. (The layout of the translated textbook page follows the original. Reprinted with the permission of the copyright holder.)

The two photographs in figure 8.5 can be thought of as showing different moments during an experiment. To uncover the intended sense, the photograph on the left has to be seen as a green empty plastic bottle with a yellow balloon, which has undergone steps 1 and 2 of the procedure articulated above the photographs; the photograph on the right has to be seen as showing the result of step 3. Our description, 'the photograph has to be seen' describes the work of structuring the ink plots on the page in a particular way; this work is indicated as structuring ('st3') in figure 8.6. Because there is no pointer (index) in the text (experimental procedure), readers have to construct (create) a link between text and corresponding photographs if they are to understand what the textbook authors intended. This constitutes unassisted, additional reading work. Before readers perform the experiment, they may identify only the green plastic bottles with (flat or inflated) yellow balloons (st3).

The inscription in figure 8.6 consists of three parts: the bottle on the left, red arrow with letter heating, and the bottle on the right dipped into water with a lit burner underneath. The two bottles have to be seen as part of a narrative in which the same bottle initially represented on its own is then dipped into the water (st1). The reader not only needs to perceive a change in the balloon, but also causally relate it to another change—the bottle on its own and the bottle in the water bath. The bottles (and the balloons tied to their openings) are depicted in cross-sectional perspective, onto which little entities with tails are drawn (st1). The lines making the bottle's walls begin with a black coloring that teeters off in a gray-white gradient to aid in perceiving the bottle in a three-dimensional form (st1). The balloon over the bottle on the right has been drawn with an orange-white gradient, assisting readers to perceive it as a representation of a balloon (st1). Only after this structuring work has been done can a link be established to the photographs in figure 8.5: we denote the establishment of such a link as the result of translation work 'tl4' (figure 8.7).

Each bottle has three kinds of molecular models differing in colors and in the number of circles: five light orange molecules represented by two jointed circles, two violet molecules represented by two circles, and one molecule represented by two violet circles with a green circle in the middle. All circles that constitute the molecules have similar color gradients, a convention in the arts to create a three-dimension impression of two-dimensional images. Such conventions, however, are not natural; they cannot be inferred from the inscription but have to be acquired within a community of users. The movement of a molecule is insinuated by semi-circles following the circular entities—this, too, is a convention that has to be learned in a community of users. The distance between the semi-circles changes, leading to their gradual disappearance: the line of the semicircles has to be articulated as showing the trajectory of the entity ('molecule') to which it is attached (st1). Initiate readers, that is, those who already know the particulate model of matter, will note the difference of these semicircles between the left and the right bottles: the intervals among semicircles are larger in the molecules in the right bottle, making the whole length of the assemblage longer (st1). Why is this difference drawn as such, and what this is for? Readers have to resort to another part of the page, which is another sign complex, such as the caption and the main text to make sense of this perceptual 'discovery'.

The main text has to be understood as suggesting that the movement of molecules becomes active, and the caption reads that it becomes faster—here 'has to be understood' means a particular kind of structuring work, which we denote as 'st2'. This statement has to be related to the difference in the images

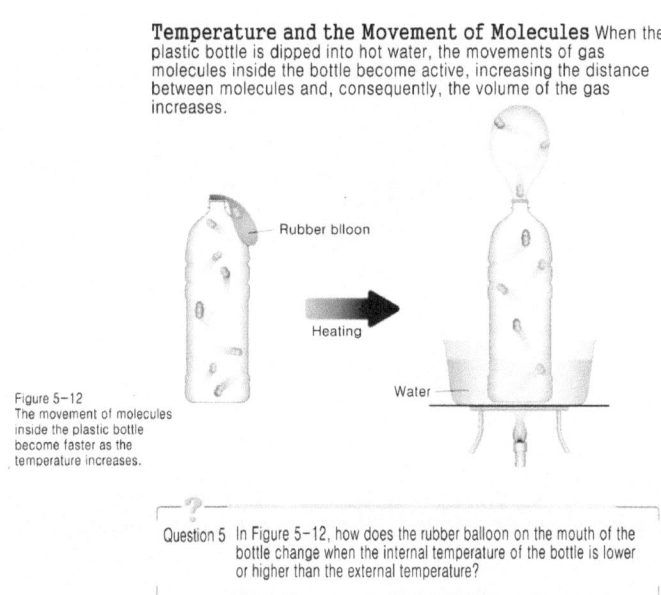

Temperature and the Movement of Molecules When the plastic bottle is dipped into hot water, the movements of gas molecules inside the bottle become active, increasing the distance between molecules and, consequently, the volume of the gas increases.

Rubber biloon

Heating

Water

Figure 5–12
The movement of molecules inside the plastic bottle become faster as the temperature increases.

Question 5 In Figure 5–12, how does the rubber balloon on the mouth of the bottle change when the internal temperature of the bottle is lower or higher than the external temperature?

Figure 8.6. The main text, molecular inscriptions, and question following figure 8.5. The inscription is placed within the main text, and therefore becomes part of it rather than constituting an add-on. The subsequent 'Question 5', also part of the main text, points students to the figure as a resource for answering. (The layout of the translated textbook page follows the original. Reprinted with the permission of the copyright holder.)

of trajectories of molecules between the left and the right bottles, for the figure to be interpreted as intended (tl5). However, the lengths of the conglomerates do not *show* the movements, inherently, because the figure itself is static. Readers have to construct or reconstruct the dynamic aspects from static images (tl1). That is, while the text *tells* us something *about* movement, the figure does not *show* movement: here, there is a contradiction between text and figure. The non-initiate reader will come face to face with this contradiction during the work involved in translation tl5.

Reading an inscription that is intended to represent particles includes special presuppositions that are related to the very lesson what the inscription is about to say. We seldom encounter inscriptions that show a microscopic world (particles) except in chemistry textbooks. Such inscriptions, therefore, resort to specific features, conventions, or assumptions shared by scientists. To see atoms or molecules, for example, one might use a very powerful magnifying glass,

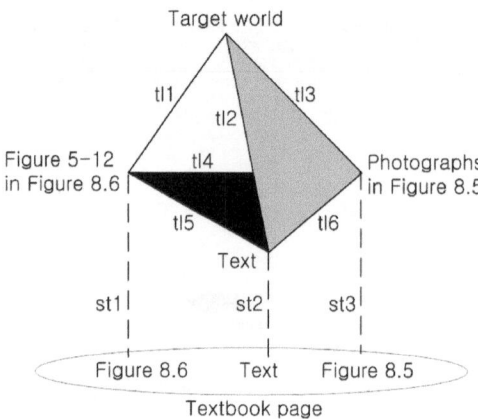

Figure 8.7. Different types of work required for reading inscriptions and texts in figures 8.5 and 8.6. Each line represents the work of structuring ([st], dotted lines), or translating ([tl], black lines).

which is the important assumption underlying the inscription. Another contradiction appears within the inscription, the drawing of the two bottles (figure 8.6). The representations of molecules, aspects of the particulate model, are depicted inside the representation of the plastic bottle, an aspect of the macroscopic world accessible to the student. Therefore, an iconic representation of a macroscopic phenomenon (bottle) is conflated with the representation of a model—two very different phenomena at the epistemic level, a perception on the one hand and a model on the other (contradiction in st1 and tl4). To understand the difference, readers have to perceive the same image as being composed of two different layers, a cross-sectional (macroscopic) layer onto which a second, microscopic model layer is superposed. Readers have to imagine seeing the entities (molecules) that pertain to the model as moving inside a bottle (tl1).

   This imaginative work requires several presuppositions. Eight molecules (from the model) drawn in each bottle (from the macroscopic world) have to be thought as exemplifying typical air molecules (st1). In this case, the author might have drawn the nitrogen gas (five light orange molecules), the oxygen gas (two violet molecules), and the water gas (one green-violet molecule); there is no explanation about the different kinds of molecular models in that number. However, these eight entities simply have to be regarded as molecules; it is not necessary to distinguish different kinds of molecules, nor is it necessary to relate this difference to any other property such as the velocity of movement. Readers need not or ought not attend to the difference (kinds of molecules); rather, they ought to concentrate their attention on other features, that is, the distance among molecules, the length of trajectories, and so on (st1). That is, the non-initiate is asked to attend to some differences but not to others. But to decide which differences to attend to and which ones to neglect, readers need to know what the lesson in the text is intended to teach. The expression 'the movement . . . increases', which appears both in main text and caption (figure 8.6), intrinsic motion is attributed to the molecules even before they are heated (st2). This intrinsic motion, which is a difficult form of discourse for students, can be reconstructed from the drawing of the bottle on the left (figure 8.6) if the reader knows the conventions for representing motion by means of tails (st1 and tl4). In addition, the molecules are scattered evenly inside a bottle so that talk about the distance between any two molecules represents the same distance. Even though the distances between molecules are not the same in this figure, they are assumed to 'be the same' or they are averaged into one (same) pictorial value (not necessarily in numerical value) (st1).

After readers accept (at least some of) these assumptions, they can relate the 'different movements of molecules between the two bottles' to 'the intervals or the whole length of semi-circular lines trailing each entity', and to 'the different distances between molecules in the two bottles'; this *relating* work is denoted in figure 8.7 by 'tl5'. Should a reader not assume the molecules are evenly distributed in any enclosed space, all the molecules *could* go upward into the balloon when the bottle is heated, and therefore reproduce a commonly reported student conception pertaining to the phenomenon (then, the distance among molecules would have to decrease).

Insufficient semiotic resource in the textbook may invert causes and effects. Another confusing factor arises in a form of the inversion of cause-effect relationships. In the present situation, the main text describes the increase in volume as the result of the increase in the velocity of molecules and in the distance between molecules. In figure 8.6, however, these scientifically sequential (logical) mechanisms cannot be easily authenticated. Rather, it is easy to construct the opposite sequence. That is, it is possible that the *distance* between molecules increases because the *volume* of gas increased. Figure 8.6 presents only the starting and ending points of the heating process, which has to be imagined. Without intermediate stages (e.g., provided by a computer animation) portraying a mechanism of how molecular movements give rise to the increase of volume, the cause-effect relationship may appear as a contradiction between the main text and the inscription (contradiction in tl5).

The label 'Question 5' associated with the drawings (figure 8.6) also leads to potential contradictions. The question refers to internal and external temperatures. However, the drawing depicts an experimental situation in which the bottle is dipped into hot water, without mentioning the temperature outside the bottle; furthermore, the external temperature differs in and above the water. Readers can compare the temperature inside the bottle only before and after the dipping action (st1). Responding to the question, then, means finding which bottle is in a state that the temperature inside the bottle is lower or higher than the temperature outside (tl5). Yet even when the right bottle is assumed to have a lower or higher temperature, there is no sign available in the inscription that the left bottle actually represents the former case. Whereas the figure depicts a static situation, the question asks about a dynamic situation; when the temperature inside the bottle is lower (or higher) than the outside, the volume will decrease (or increase), which cannot be authenticated by the reader in the depiction (contradiction in tl5).

Macroscopic activity

There is a huge, potentially insurmountable gap between a role-play (inscription) and the microscopic world for which it provides a model. Role-play, as a type of activity, can provide an opportunity to experience situations from another perspective. However, most role-play strategies in teaching science are applied mainly to the social (environmental) or historical situation of humans. Can a microscopic world be role-played by students? New kinds of role-play, as we show here, in which students behave like molecules, have more limitations than advantages. Take a look at figure 8.8, which shows the way in which students have to enact a particular activity. In this role-play students become molecules. The aim of the role-play is to let students experience the way in which molecules move to explain how air lifts a car with a flat tire. The author may have intended letting students feel how molecules exert forces while colliding with the container. However, we may ask, what sense can students make of this activity (inscription) during and after role-play?

The inscription does not depict a model of the microscopic world as other conventional molecular models do. Nor does it show the (molecular) movement: it is a static image. Students enacting different roles will experience different things. Only the 'molecule' students can experience the movement of molecules. The students role-playing the walls are asked to stand still even after the molecule students bump into them. Many assumptions or analogies have to be unearthed to link this activity to the microscopic perspective. The two-dimensional square has to be thought of as corresponding to a three-dimensional space (tl3 in

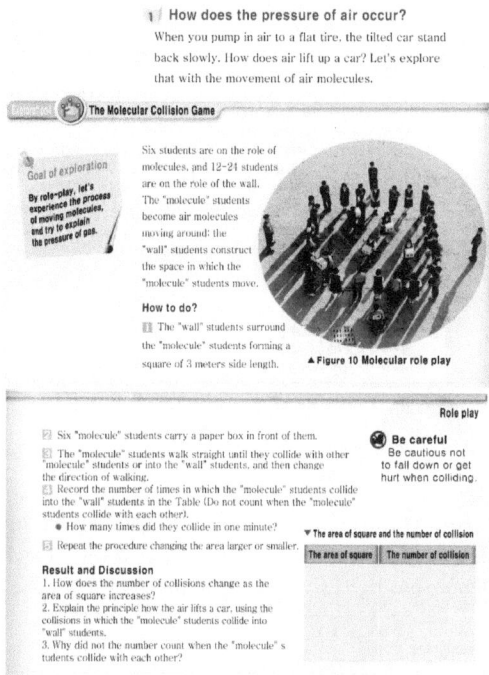

Figure 8.8. The macroscopic activity using role-play. (The layout of the translated text-book page follows the original. Reprinted with the permission of the copyright holder.)

figure 8.9): in realistic situations any such two-dimensional square at molecular level is hard to get (baring nanochemistry). In addition, the square has to represent a closed system in which the number of molecules is fixed. The molecule students are imaginary (representative) entities on the inside of the closed system. The differences of students (such as, male or female, tall or small, etc.) have to be disregarded (tl3). In addition, students have to avoid the pitfalls of anthropomorphization: the molecules do not behave in the same way as living human beings, for example, they never stop moving.

The situation of role-play is different from the flat-tire situation that students have to solve after this activity. Whereas the area in the role-play is a kind of closed system, the flat-tire situation is not strictly closed because more air is pumped in (contradiction in tl6). Considering the goal of the exploration in figure 8.8, it is more plausible to increase the number of molecule students instead of changing the area of the square, because the relationship between the area of

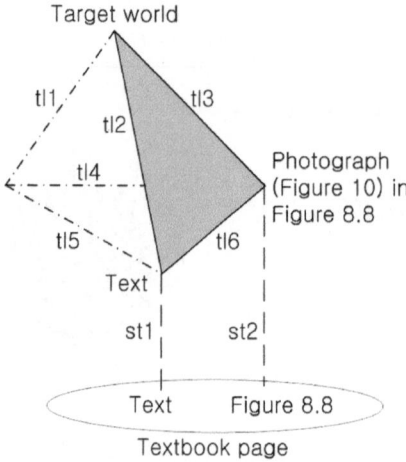

Figure 8.9. Types of work required for reading inscriptions and texts in figure 8.8. Each line represents the work of structuring ([st], dotted lines), or translating ([tl], black lines). The one-dot-chain line represents the missing line from the semiotic model in figure 8.4.

the square and the number of collisions correspond to Boyle's law, which is not the relevant concept in this activity.

In general, the text does not clarify why this role-play is needed, why students take the roles of the 'molecule' or the 'wall'. Students have no prior experience of being a molecule, and then, what is intended by experiencing molecules? Every action (walking and colliding), which itself is a new kind of experience, has to be translated into new, microscopic phenomena. For example, while molecule students are walking, they have to think that they are like an air molecule moving in certain direction, and that they will exert force by colliding only into students representing the wall (tl6). The collision experience will be the same whether they collide with the wall or other molecules. Here a critical contradiction arises: why do some collisions not contribute to the pressure while they are felt the same? The third discussion item in figure 8.8 asks students to solve this contradiction (st1, st2). However, the role-play activity does not guarantee students' understanding of this discussion item. There is no motive for the activity in this kind of cookbook style role-play.

Even after molecule students experience collisions and link those experiences to the molecular collisions analogically, they may not understand the concept of pressure because an *individual* molecule does not explain gas phenomena. Pressure is a *collective* effect of all molecules colliding with the wall of the container. Students participating in this role-play have to think collectively not individually, if they are to construct the appropriate meaning. That is, after the role-play finish and the table (in figure 8.8) is filled with numbers, all students can conjecture the number as representing the pressure (tl2). However, the collision of students in the role-play is still quite a different phenomenon from the microscopic collision of molecules in a container. One molecule student might think, 'I can collide into another molecule student and change the direction so that I can go to the wall student more quickly, or more slowly. Then, how can I link the effect of other molecule students to the pressure in total?' In this sense, this role-play is dangerous from a didactic point of view.

This role-play can hardly be an analogy. In order for this to be an analogy, students already need to know what an analogy is. Because both the actions (walking or colliding) and the meaning of the actions (in microscopic perspective) are entirely new or unfamiliar things to the students, the role-play is potentially something that distracts the actors from learning the target concept (pressure). Anthropomorphic explanation (or experience) of molecular movement may lead to students' misinterpretations, because the (role-playing) students are required to execute a kind of schizophrenic process. All these problems that this role-play possibly generate can be shown in our model (figure 8.9). That is, students have to make big strides to bridge the gap between their experience in role-play and the microscopic world.

Anthropomorphic shapes in arbitrary space

Surface features of inscriptions (such as color, shape and background) may restrict the reading process. Almost all inscriptions in Korean textbooks are presented in color. Any colorful inscription may catch readers' attention at first, but can also confuse the reading process. Novice students likely focus on surface features (e.g., color) of inscriptions rather than on underlying concepts. Based on their prior, successful experiences of navigating their world (e.g., looking through a microscope), many students think that atoms, too, have macrolevel properties such as various shape, color, or smell. In science learning, the instructional material thereby becomes a potential source of students' conceptual difficulties or (mis-) conceptions. The chemical inscriptions encountered in textbook will mediate understanding chemical concept in ways unintended and undesired by teachers.

The inscription in figure 8.10 is an example of an illustrative inscription, located in an arbitrary space (background) using anthropomorphic shape as a basis of a molecular model. Each model (yellow circle) has anthropomorphic characters with similar (not exactly the same) blue color. Each character has a triangular face (or body) with eyes, mouth, arms, and legs similar to characteristics that may be found in commercial cartoons (st1). The left circle has nine characters holding each other hand-in-hand; the middle circle has eight characters sometimes in hand-in-hand manner or sometimes alone, walking or turning back; the right circle has three characters all alone, running or jumping (st1 in figure 8.11). How can readers determine which model corresponds to one of the three states of matter, the purpose of this exploration? What is needed to learn by interpreting this inscription for understanding?

Reading this inscription may be made difficult because of several inherent contradictions. The first contradiction arises from the fact that the (static) figure does not show movement whereas the caption and the main text describe the situation with the term movement (contradiction in tl5). We already used the conventional signs in reading the movement in the three circles: especially in cartoons, the authors use special signs, such as double parentheses (in the middle and right circles) or a white cloud-like tail (on the right circle), to express the movement of the character. Even though all readers are assumed to be familiar with these conventional sign, it is not enough for readers to interpret what each circle depicts and therefore to correspond to some physical state. The index 'Figure 7-7' (figure 8.10) only shows the movements of blue characters that cannot be linked directly to the state of matter (additional reading work is required as we below). In addition, while the main text reads, 'molecules are moving ceaselessly', the characters on the left circle do not have any conventional sign depicting the movements other than some facial expression or the curved strings that can be seen as representing hair (tl5). The left inscription hampers the interpretation of *movement*, because it does not show the phenomenon itself.

Another contradiction arises with the arbitrary space (background) where the models are located. This space, here the yellow circles, does not give any indication as to what it depicts. Does a yellow circle represent the cross-sectional space in the air or the virtual playing ground where the characters can move around? Without such assumptions the circles cannot be compared. Initiate readers can use other conventional tools to unravel the physical reference of the yellow circle, focusing on the four characters on top of the left circle, which are smaller than other characters; the circles therefore depict something like a playground (st1). The classical artistic and photographic conventions couple the distinctions high versus low and far versus near. However, a non-initiate may not have the semiotic resources to infer what is happening in the three circles such as, 'What is the movement like?' 'Why are the numbers of characters different?'

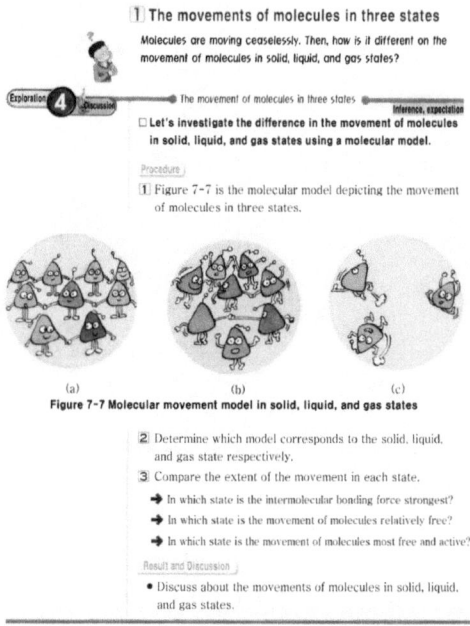

Figure 8.10. The example of molecular model using persons in arbitrary spaces. (The layout of the translated textbook page follows the original. Reprinted with permission of the copyright holder.)

'Is the yellow space a ground or in mid-air where the characters fall from an airplane?' 'Is the movement of molecule similar to the bodily movement of each character?'

An internal comparison of figure 8.10 also yields a contradiction. The first question following the third step of the procedure asks about the intermolecular force. However, the third step requests readers to compare the extent of movements only. Until the readers come to the question, there are no terms such as force or bond. The inscription does not show the intermolecular bonding force directly. Only initiate readers can make a conjecture that the joined hands may correspond to the intermolecular bonding (tl5). The figure does not differentiate which holding is stronger than the others. The number of characters holding hands differs among the circles: all nine (or more) characters are holding hands together in the left circle; six characters are holding hands in the middle; no characters are holding hands in the right. These numbers of holding hands then

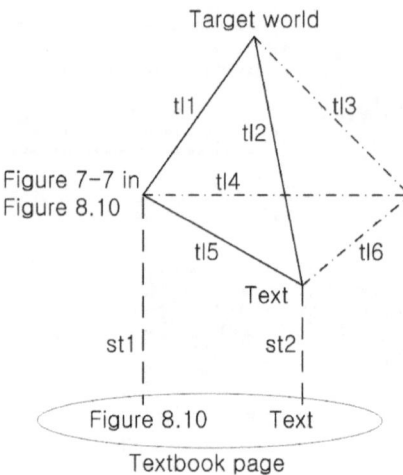

Figure 8.11. Different types of work required for reading inscriptions and texts in (a). Each line represents the work of structuring ([st], dotted lines), or translating ([tl], black lines). The one-dot-chain line represents the missing line from the semiotic model in figure 8.4.

have to be translated as representing a smaller number of strongly bonded entities (tl5). Could the right circle—where there are no holding hands—be thought of as if there was no intermolecular force? There is sufficient ambiguity so that students cannot consider interactions between particles.

Although the text and the caption use the word 'model', the nature of this model is not articulated. If readers think the model as a copy (replica) of concrete object, they might construct the image of molecule in a form of a person depicted in this inscription. If the purpose of this inscription is to show the difference of molecular movements in three states of matter, the author could have used simple shape (such as, for example, circles) to represent a molecule. To interpret the molecular model in figure 8.10, readers have to ignore certain features of inscription, such as the difference in the colors, shape of mouths or eyes, or size of characters as irrelevant aspects of the model. Otherwise, this inscription could provide an avenue for the breeding of alternative conceptions, restricting the reading process.

Circles and worksheet in magnified space

Constructing an inscription is a dialectic process of making assumptions and learning a chemical concept. Scientists make chains of inscriptions by translating world (matter) into other forms of sign or language. In doing so, scientists employ various social conventions. In previous sections, we show the work of translation between inscriptions requires discourses related to the underlying concepts. In a similar way, transposing an inscription requires familiarity with the discourses concerning the underlying concept as well as with the encoding conventions and assumptions shared in the community of chemists.

Figure 8.12 contains inscriptions of illustrative and worksheet functions, located in a magnified space and using circles as the basis of the molecular model. In this activity, students first observe a macroscopic phenomenon of volume change in an experiment of heating acetone in a closed system (plastic glove). Then the textbook provides a molecular model of acetone in the liquid state, followed by an empty space where students have to draw on their own models of acetone in gas state. The stated purpose is to allow student to understand the change of state from liquid to gas, the change of the volume, in terms of different molecular arrangement. The main text following the title 'Exploration 10' provides an explanation of change of volume and state.

The inscription is comprised of two or more genres of inscriptions: photograph and molecular diagram (left side) and schematic drawing (right side). Pale brown lines and shadows between the small circle inside the left photograph and the molecular diagram can be perceived as a magnification of the liquid acetone in the thumb of the plastic glove (tl4 in figure 8.13). The first step of the procedure ('Put some amount of acetone into. . .') has to be linked to this left part of the inscription (tl5); the caption figure 24 (a) (in figure 8.12) is indexed in the text (result and discussion 1) as 'The molecular model of acetone in liquid state is like (a) in figure 24'. The terms liquid or acetone in the text and caption are potential resources in this translation work (tl5). The blue arrow between the two photographs has to be perceived to denote the process of heating in the experiment, although neither the text nor the inscription provide such explanation. On the right side, the schematic drawing of the plastic glove, which looks like the cross-section of a glove partly overlaps the photograph of the inflated (this also has to be perceived as such; st4) glove. With reference to the molecular diagram provided, students are requested to draw acetone molecules in the gaseous state (tp2). Thus, the right part of inscription should be thought of as depicting the final result of the second step of the procedure (tl12).

2/ What is the difference in the arrangement of molecule according to the state of matter?

Let's investigate how the arrangement of the molecules becomes different when the state of matter changes, with following activity.

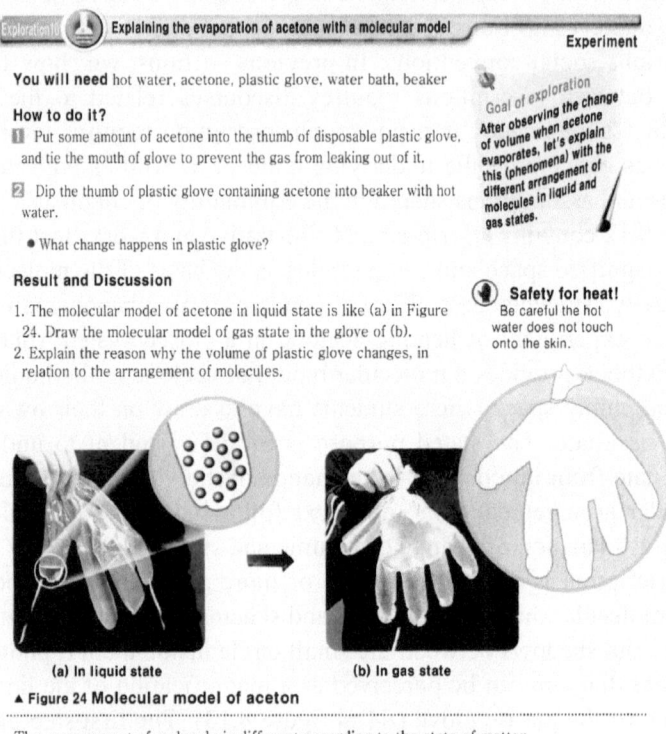

Exploration10 Explaining the evaporation of acetone with a molecular model

Experiment

You will need hot water, acetone, plastic glove, water bath, beaker

How to do it?

1 Put some amount of acetone into the thumb of disposable plastic glove, and tie the mouth of glove to prevent the gas from leaking out of it.

2 Dip the thumb of plastic glove containing acetone into beaker with hot water.

• What change happens in plastic glove?

Result and Discussion

1. The molecular model of acetone in liquid state is like (a) in Figure 24. Draw the molecular model of gas state in the glove of (b).

2. Explain the reason why the volume of plastic glove changes, in relation to the arrangement of molecules.

Goal of exploration
After observing the change of volume when acetone evaporates, let's explain this (phenomena) with the different arrangement of molecules in liquid and gas states.

Safety for heat!
Be careful the hot water does not touch onto the skin.

(a) In liquid state   (b) In gas state

▲ Figure 24 **Molecular model of aceton**

The arrangement of molecule is different according to the state of matter. For this, the volumn changes when the state changes. The distance between molecules of gas state is more longer than those of liquid or solid state, the volume of gas increase very much.

Figure 8.12. Example of inscription with an empty space where the reader should draw molecules. (The layout of the translated textbook page follows the original. Reprinted with the permission of the copyright holder.)

Before and during the process of drawing, students have to make assumptions in their progress of interpreting and thinking. They have to assume that the fifteen blue circles (or balls; it does not appear to be required here to distinguish these things) represent all molecules of liquid acetone they used in the experiment (tl1); that these circles do not change their size, shape, or number while undergoing the change of state (tp1). As seen in the photograph on the right side

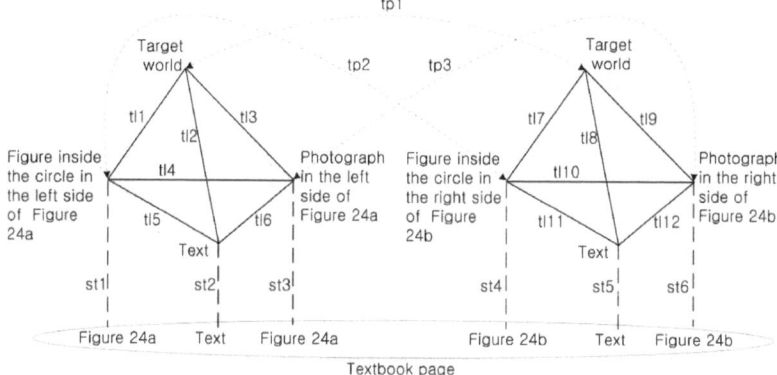

Figure 8.13. Work required for reading inscriptions and texts in figure 8.12. Each line represents the work of structuring ([st], dotted lines), transposing ([tp], two-dots-chain arrows), or translating ([tl], black lines).

and in students' experiment (or possibly in teachers' demonstrations), the acetone is colorless, but the molecular models have blue color. Thus, in gas state also, although students can see nothing inside the transparent glove, the molecules will have the same blue color. Then, the color of the molecular model could be one source of contradiction or one thing that has to be disregarded as irrelevant feature of the model in this activity (contradiction in st1).

The different scale of the schematic drawing from the molecular diagram could—if perceived (processes st1 and st4)—also give rise to confusion (contradiction). Students may see the acetone molecules of *such* size in the left diagram (only the thumb of glove is enlarged). Then, students might ask themselves whether they should draw the gas molecules *smaller* in size on the empty glove because the entire glove is enlarged? When students perceive the different ratio of gloves and drawing so that they draw molecules in different sizes, this change in size may contradict the scientific concept and the assumption that the molecules are the same regardless of their states (contradiction in tp2). Or, students might draw larger gas molecules, because they have reported an incorrect notion that the sizes of particles increase with change of state. To resolve these two contradictory situations (regarding color and size of the model), students are required to already understand the chemical concept.

The process of drawing and its results do not guarantee finding the reason why the plastic glove is inflated. There is a gap between the cause and the effect of a phenomenon. Students are expected to draw molecules in *evenly distributed*

manner across the given space (st4), which is another important assumption
shared in scientific community. The sequence of this activity seems to assume
that, when students draw molecules evenly in the inflated glove, they will come
to the understanding of the macroscopic phenomenon in microscopic perspective, in which the gas molecules of acetone have to be distributed evenly, *resulting in* the change of volume. Interpreting the macroscopic phenomena in this
way requires that students have prior experience or scientific knowledge *already*: for example, in order to change the shape of something, one must exert
forces on that thing; the gas molecules (balls) tend to disperse in all directions
(the increase of entropy); or the balls (molecular model) exert forces when they
collide into something. Without understanding or discussion of these prerequisites students are forced to draw something in the text as a procedure of this
cookbook style activity. They possible may conclude that *because* the volume
increases the acetone molecules in gas state *can* distribute evenly. Or, there is no
way to dispute this conclusion with the drawing (activity) itself. To understand
the cause-effect relationship, students are required to have the relevant knowledge beforehand. Only after students finish the activity, the main text provides
an explanation on the change of the volume. That is, constructing (transforming)
an inscription and understanding the intended lesson presuppose another. Students need to know the underlying concept and assumptions (conventions) to
transform an inscription; and they need to construct inscription to understand the
lesson.

Unraveling the contradictions

Previous research showed that students experience difficulties in understanding
and learning the particulate theory of matter. In response to these difficulties,
Korean textbook authors have used large numbers of diverse inscriptions intended to assist students in bridging the gap between their understanding of the
macroscopic world and the particle-based models that are used in explaining it.
Our chemisemiotic analysis revealed the tremendous amount of work involved
in learning from the inscriptions, work that may be far beyond what any chemistry novice brings, let alone the seventh-grade students targeted with the inscriptions in our investigations.

     Our analysis shows that reading inscriptions and texts require several kinds
of work: structuring each inscription and text (caption), transposing inscriptions,
linking (translating) inscription with text, and interpreting the meaning of them
(linking the textbook with target world). Thus, we propose the unit of analysis
should include all inscriptions, captions and main text. These elements in the

textbook constitute a segmentation of the material continuum that constitutes this world. In the reading thus interpreting these elements, we can cast doubt on the way in which the continuum has been segmented. Our model makes salient different genres of signs; these have to be combined to denote the target world, another segmentation of the material continuum. The three elements on the bottom plane in figure 8.3 constitute the background of each other, contributing to their respective sense.

We identify many contradictions in the reading work required to uncover the intended sense of inscriptions toward a standard chemical discourse. Contradictions are sometimes inherent in a sign itself. The space in which the molecular models are drawn is one source of contradiction including the layered space in figure 8.6 and the arbitrary space in figure 8.10. Another source of contradiction exists in the way the models are expressed; models have irrelevant features such as colors, number of circles (in figure 8.6), and shape of characters (in figure 8.10). More frequently, contradictions occur in translating different signs into one another. In this situation, while the text articulates the movement of molecular model, the inscription never shows it. Transposing work may contain contradictions as in figure 8.12. In addition, the cause–effect contradiction in figure 8.6 can also be problematic.

How can these contradictions be resolved? In the case that the contradiction emerges from the structuring work, the reader cannot resort to another sign to solve the problem. For example, when there is no explanation what the arbitrary space means, readers can only conjecture as to its sense, sometimes resorting to the social conventions of drawings. Therefore, sufficient resources have to be provided so that students can make sense of these kinds of inscriptions and thereby develop the graphicacy required in critically producing and using them.

Another case of contradictions can be solved partly because of the way translation works. Thus, different elements that embody the same sense are not only contradictory but also complementary to each other. Although the textbook image itself cannot show the movement, the subsequent caption and text are supplements that provide the term movement. Readers then can recognize a certain feature (such as tails or double parentheses) as depicting the movement of the molecular model. That is, the text supports or mediates the reading of the inscription. In this way, an inscription and a text can be read in a triadic relation (the face on the pyramid of our semiotic model) to the third element, target world.

Many ontological and epistemological issues pertaining to the nature of a chemical model remain even after our analysis. Students may see different things in an inscription that is intended to represent particles, especially when they learn the particle model first. There are only red balls (in figure 8.1) or triangular characters (in figure 8.10). The balls or characters are not given as tools for explaining the phenomena. Students' difficulties in learning the particle concept may be inherent in the inscriptions or may be due to the lack of experience with the discourse about particles rather than due to cognitive deficiencies. To develop (critical) graphicacy, students will need practical experience in using modeling discourse if they are to learn predicting and explaining observed phenomena. Naming and using a model in concernful activity will provide students with the ontological character of the model.

One of the fundamental pillars of chemistry—the theory of the particulate nature of matter—has been a notorious problem in chemical education. There appears to be a gap between students' experiences of and in a macroscopic world and the particle-based explanations of the model used to explain the former. In the past, researchers attributed students' difficulties to a lack of knowledge, misconceptions, or mental deficits. However, recent work in science education and language make assertions that such deficit approaches have limit to explain the *practice* of using language and inscriptions. Competent reading activity and critical graphicacy are therefore possible only when readers are familiar with language, inscriptions, and the referents to which they pertain. A chemisemiotic analysis of reading materials allows us to understand where readers' (students') difficulties may emerge.

9   Reading layered, dynamic inscriptions

Throughout this book so far, we analyze inscriptions that appear in textbooks in terms of the work required to read them according to the authors' intentions; we consider both simple and layered inscriptions. We also analyze the graphicacy that students of various ages (eighth through eleventh grade) exhibit when they read, interpret, and transform inscriptions. So far, however, we have not considered the opportunities computers provide users to exhibit graphicacy, that is, manipulate inscriptions, which are then animated by the medium for exploring the consequences of the manipulation. In this chapter, we develop issues in the semiotic and anthropological analysis of inscriptions that we articulate in previous chapters: layered inscriptions (chapters 7, 8) and the salience of signs with movement and gesture during face-to-face interactions (chapter 5). That is, we provide analyses of (a) the inscriptions that appear as part of a software program used in physics courses as a means to model movement phenomena and (b) student-student and student-teacher interactions over and about the inscriptions and events on the computer screen.

Layered, dynamic medium

Our analyses of the work of reading layered inscriptions (chapters 7 and 8) show that the work seems to become easier as layered inscriptions can provide transitional steps between very concrete photographs, where students see and recognize things they are familiar with, to very abstract inscriptions that have little similarity with the things they stand for, such as graphs or equations. The work also becomes more complex, because layering means having several inscriptions appear on the same part of the book page: perceptual work has to be done to structure *one* part into *different* inscriptions, and then to relate the resulting inscriptions or parts thereof according to the underlying metaphor, for example, the iconic relation between the height of a point on a graph and the height of an air-compressing piston above its bottom (see figure 7.1). Common to all of the inscriptions we analyze elsewhere in the book is the fact that they are static and that any dynamic relation has to be read into them. Computers provide the opportunity to make images dynamic, giving rise to new phenomena heretofore not considered. How do students read and interpret layered inscriptions when these are rendered dynamic? That is, what are the levels of graphicacy displayed in this relatively recent form of inscriptions? Despite the hype that the computer revolution generally and the more recent coming of the Internet more specifically have caused among educators, there exists very little good research and thorough analysis of how these media have changed the way students interact with inscriptions generally and with animated inscriptions more specifically. Here we take a look at one piece of software widely used by physics teachers since the early 1990s because of its apparent ease of use.

The medium

While one of the authors (Roth) was teaching fulltime high school physics, he used a computer-simulation software to provide his students with opportunities to simulate motion phenomena using *Interactive Physics™*, a piece of computer software that simultaneously displays representations of phenomena and conceptual entities (figure 9.1). *Interactive Physics™* (here running on a Macintosh platform) is a computer-based Newtonian microworld that allows users to conduct motion-related experiments. Observables (measurable quantities) such as force, velocity or acceleration can be represented by means of vectors or instruments such as strip chart recorders and digital and analog meters. All student activities included, at a minimum, one circular object (figure 9.1). A force (full arrow, shaded in figure 9.1) could be attached to this object by highlighting and moving it with the mouse. The object's velocity was always displayed as a vector; the students could set its initial values, length (i.e., speed) and direction. At the outset, students were not told what the arrows are standing for. Students were asked to find out more about the microworld, especially the role of the arrows and their relation to the events that were displayed on the screen. Some of the teacher-designed activities displayed nothing more than the circular object, its velocity, and a force. Others required students to manipulate 'the arrows' (force and velocity) to hit a small rectangle and to throw it off its pedestal. After setting force and initial velocity, students could execute the experiment by hitting the 'run' button (top left of the window). A tracking feature 'froze' the motion as if recorded with flash photography (figure 9.1). During a simulation, the cursor takes the form of a stop sign; a simple mouse click stops the motion. The replay feature (bottom left of the screen in figure 9.1) allows the inspection of individual states in the motion of the sphere. Before completing the assigned ac-

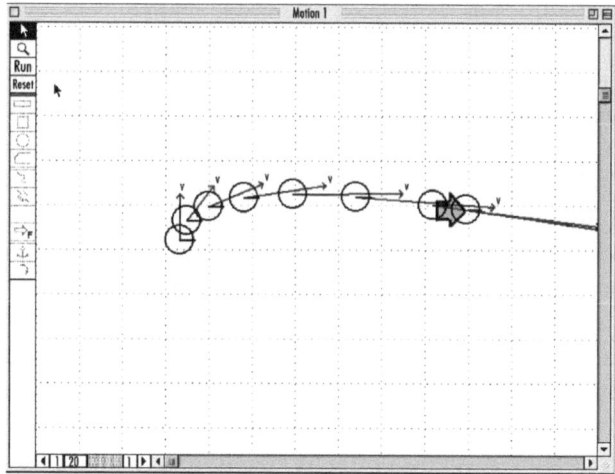

Figure 9.1. The students see this interface, which consists of the objects that they can manipulate, here the circular object and the two arrows, and tools that surround the center area, in particular the buttons that start ('Run') and reset a simulation. During a simulation, the cursor turns into a stop sign and a simple mouse click stops the event. There is no inherent indication that the circular object and the two arrows really depict very different kinds of entities.

tivities at their own pace, students were introduced to the key features of the interface during a twenty-minute demonstration; additional operating instructions were provided with the instructions for the activities.

Semiotic analysis

The computer display can be segmented into a tool section surrounding an area, in which the events actually happen. The central area is clearly distinguished from the remainder of the interface in several ways. First, on top, right, and bottom, the central area, appearing in white with a dotted grid, is set apart by the grey color of the border. On the left hand side, the tool section, although featured on a white background, is set apart from the central area by a solid line that interrupts the horizontal dotted gridlines. There are small box-shaped segmentations, each containing an icon.

In the central area, there are different kinds of entities, which the students are supposed to perceptually separate—i.e., structure—and learn about. These entities include arrows and various kinds of circular and rectangular objects.

There are other entities as well, such as springs and connectors (see left bar of the interface [figure 9.1]), but for simplifying our presentation, we focus on the circular object and the two differently shaped arrows. These two types of entities are inscriptions that point to different domains. The circular object and its motion belong to the 'phenomenal domain'; as such, objects and motion are part of students' everyday experience. The arrows, which an expert knows to denote velocity (single line arrow) and force (outline arrow), are entities from the 'epistemic domain'. We use the notions of 'phenomenal' and 'epistemic domains' to distinguish between a more phenomenological, holistic, and undifferentiated human experience of 'things out there', and the highly specialized discourse of the scientific community and its associated inscription practices. As part of their discourse, physicists use vector diagrams to analyze empirical events; that is, they attach to these physical concepts, numbers, and graphs. Non-physicists, however, often see little connection between discourse and the events they witness. *Interactive Physics™* was designed to bridge between the two domains by enabling what is impossible in the world of our everyday experience: the layering of the phenomenal (moving object) and the conceptual (representations of velocity and force). This copresence is achieved by transforming three-dimensional real-world objects into two-dimensional drawings—which resembles the process of re-representing a train in a photograph on which vectors can be layered, a situation analyzed in chapter 7. At the same time, the constructs of forces and velocities are transformed into drawings (vectors) in the same two-dimensional plane of the computer display. As a result, conceptual properties and their associated microworld objects are copresent, each re-representing a different aspect of human experience (the conceptual and the phenomenal).

Drawing on the semiotic framework presented in chapters 7 and 8, we can now analyze how the entities on the interface relate and the work of reading and interpreting required for understanding what they signify. In a first step, the reader has to identify the circular object, its motion, the arrows, and the changes they undergo—whether users make these distinctions is an empirical matter. That is, readers need to *perceptually structure* the interface in such a way that pertinent features become salient, at which point they can serve as signs. We denote this work as *structuring₁* (figure 9.2) the work that separates the phenomenal objects and events—the circular object and its movement characteristics—from the ground as well as the arrows (vectors). Now that vectors and circular object have become figure against the remainder of the interface as ground, they also need to be seen as different kinds of entities, the former constituting elements of a theoretical framework describing and explaining the latter. Thus, a second structuring movement is required (*structuring₂*)—which also involves a

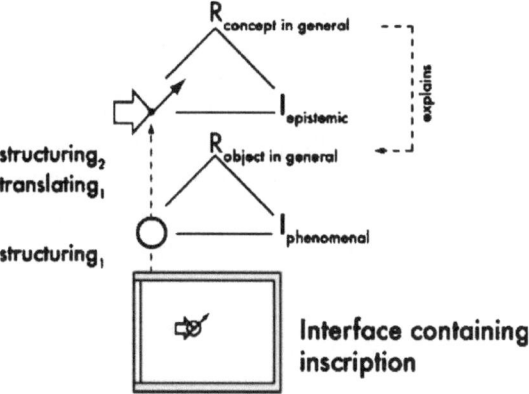

Figure 9.2. To understand the events on the interface, users need to perceive the circular object, its motion, and the different arrows as figure against a ground. They then need to further structure the display, attributing object and arrows to different domains, the latter describing and explaining the former.

translation (figure 9.2)—that separates epistemic and phenomenal entities that belong to different planes or layers.

The two different layers have interpretants and referents from different domains. Objects are related to interpretants from the phenomenal world, vectors (arrows) require interpretants from the epistemic domain. The referent of signs and interpretants from the phenomenal world is an object in general, the referent of vectors and interpretants from the epistemic domain are concepts or ideas in general. The function of concepts and ideas is to explain the objects and associated events of our experience. On the surface it may appear as if the object appeared in both the phenomenal and epistemic domains. In fact, Newtonian physics is not about extended objects but about point masses, that is, entities that have the same mass (e.g., measured in kilograms) as the extended object but have no extension, so that the mass is thought of as concentrated in one, infinitesimally small point. Once known, this point may become salient in the display. Our semiotic model includes the dot (point) at which the velocity vector starts and the force vector ends in the epistemic level rather than as 'the center' of the phenomenal object. This structuring, though it may sound evident now that we described it, may not be as evident as it appears. Thus, for many years, the first author (Roth) had thought about the dot as the center of the circular object rather

than as a point mass representing the circular object in the epistemic domain—
despite his graduate degree in physics, despite his extensive use of the software
with high school students, and despite writing a number of research reports
about student learning. It has occurred to him only now—as we conducted a
careful semiotic analysis of all the inscribed traces on the interface—to ascribe
the dot to the conceptual rather than the phenomenal world.

For the physicist, the arrows are interpretants (inscriptions) of 'vectors',
which do not have to be represented in this way. Rather, many other interpretants are used in physics practice, including those in the identities

$$\bar{v} = \begin{pmatrix} v_x \\ v_y \end{pmatrix} \text{ or } (v_x, v_y); \quad \bar{F} = \begin{pmatrix} F_x \\ F_y \end{pmatrix} \text{ or } (F_x, F_y).$$

Both the bar across the letters v (velocity) and F (force), on the one hand, and
the vertical arrangement of the horizontal (v_x, F_x) and vertical (v_y, F_y) components are different ways of denoting the same vector nature of velocity and force.
That is, from the physicist's perspective, objects, their velocities, and the forces
acting upon them are idealized, transcendental entities and therefore are inherently depicted in inaccurate fashion in real diagrams.

Although we have advanced enough from the perspective of physics, we
need to push our semiotic analysis a little further to reveal why learning physics
in general and learning with *Interactive Physics™* in particular is riddled with so
many difficulties. The issue is the unacknowledged layering that occurs as part
of physicists' modeling of motion using position, velocity, and forces. These
three are different conceptual objects and therefore require, from a mathematical
perspective, different vector spaces or grids. Let us take a look at an object that
moves along a trajectory shaped like a parabola similar to that in figure 9.1.
When the object starts, it is in a location (point) that can be described (1, 1) or
one unit in the horizontal direction and one unit in the vertical direction (figure
9.3). By convention, scientists and mathematicians use the letters 'x' and 'y' to
refer to these two dimensions; if there is a third dimension, it is named 'z'. After
one second, the object can be found at a point two units to the right and two
units up or (2, 2); after two seconds, the object is at (5, 3).

Now, the object moves like this exactly if there is a force constantly pushing the object to the right at two units per second every second; this would be
represented in force space by (2, 0) or by an arrow going two units to the right
and 0 units upward (figure 9.3). Because we said that the force is constant, it
would be the same at one second, two seconds, and so on, always (2, 0). That is,
the object continuously accelerates. If the object at the outset is going zero units

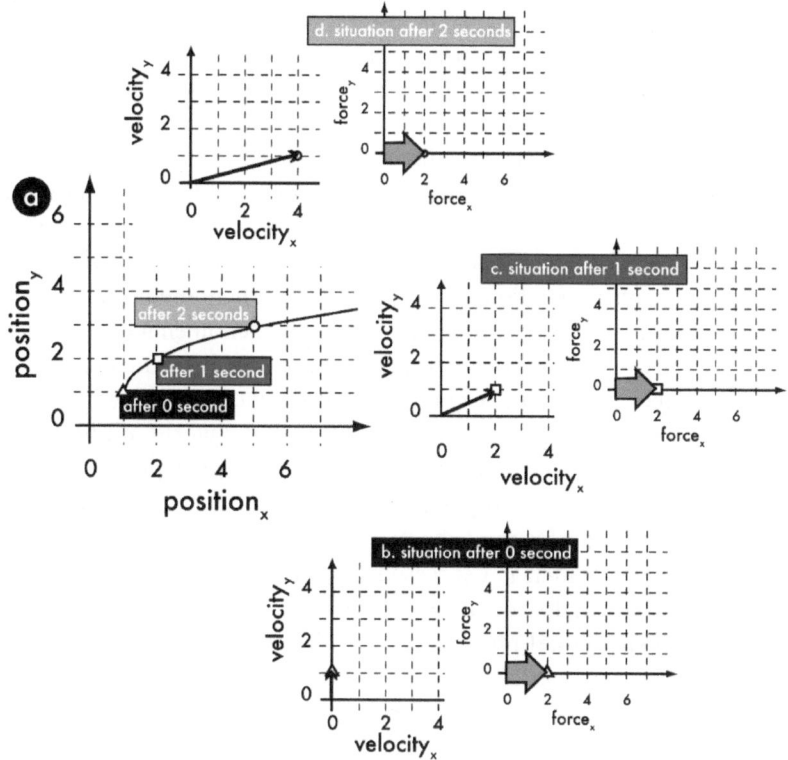

*Figure 9.3. The object in Figure 9.1 moves on a parabolic path. This is modeled here in
(a) the position graph. Three pairs of graphs show the velocities of and forces on the object after zero (b), one (c), and two seconds (d), respectively. Although the force is constant, as seen in the three force graphs, always points two units to the right (b, c, d), the
velocity changes (b, c, d). But this change is only in the velocity component to the right,
never in the upward velocity component, which always stays at one unit.*

per second to the right and one unit per second upward, this is represented in velocity space by (0, 1) or by an arrow that goes from the origin (0, 0) to the point
(0, 1). After one second, the object has a velocity of one unit up and two units to
the right, that is, (2, 1) or an arrow from the origin (0, 0) to the point (2, 1) (figure 9.3). After two seconds, the object has a velocity of four units to the right
and one unit upward, or (4, 1) (figure 9.3).

   Inspecting the three velocity panels, readers will recognize that the velocity
gets longer and directed more to the right, just as it does in *Interactive Physics*™

(figure 9.1). More so, the velocity is always (at 0, 1, and 2 seconds) one unit upward and only increases to the right. This should make sense, for the force is only pushing to the right and not upward, so the object accelerates (gets faster) only to the right but not upward.

In our presentation, there are three grids, one for each position, velocity, and force. In the Newtonian microworld *Interactive Physics™*, however, the three grids—and therefore the three concepts—are all shown in the same plane. That is, already in the conceptual world, we have a layering of three different spaces. The first space, position, overlaps with the phenomenal space of the object. That is, by placing a mathematical grid over phenomenal space, we mathematize it. That is, we have to do something for the different positions of the phenomenal object to become a mathematical object. Epistemologically, therefore, the two spaces are different things.[1] Although the world of phenomena and its description and explanation in and with theoretical concepts are two different realms from an epistemological perspective, in science they are often treated as identical. Thus, the sciences treat the world and its mathematical description as structurally identical, which is expressed in an isomorphism or correspondence of the form {fundamental structure ↔ mathematical structure}. Thus, the structure underlying experiential phenomena and the structure underlying scientists' mathematical equations are taken to be the same. The computer software, which conflates what we articulated as two levels into one, is therefore consistent with the sciences, where measurements correspond precisely with their objective referents.

In the next step, we can layer two more mathematical spaces on top of the first: in one space, we represent the velocities that the object takes in the course of its journey; in the other space, we represent the force (figure 9.3). In the physicists' way of depicting the motion of objects, the three spaces are layered. Once seen together, we get exactly the inscription displayed by *Interactive Physics™* (figure 9.1), which therefore consists of four conceptually different layers. To the novice, however, there is only one monitor surface. It is through their actions that novice users need to establish the relations between the different entities. The notorious difficulties students have in understanding Newtonian physics conceptually may lie in part in the fact that physics instruction almost never articulates the epistemologically different layers but fuses them into one, the phenomenal space.

Perceiving the sign

The purpose of the *Interactive Physics™* software is for students to learn the physics of motion. In contrast to reading a physics textbook, which contains texts, equations, and various other forms of inscriptions (e.g., equations such as $\underline{x}(t) = \underline{x}(0) + \underline{v} \cdot t$) all of which are static, the software provides students with the possibility to manipulate entities and to test the consequences of their actions by running simulations. Nevertheless, to learn anything, students need to identify salient entities that can become signs for which they produce interpretants. It is in the production of interpretants that students articulate previously unarticulated understanding and elaborate what they newly articulated into new understanding; both processes constitute elements of learning. The resultant interpretant-sign relations are grafted on and elaborate the inaccessible sign-referent relation.[2] The interpretant-sign relation is open in the sense that there is always another interpretant capable not only of mediating the relation between sign and object but also that between sign and the first interpretant. The interpretant therefore is something like a commentary, a definition, or a gloss that elaborates all previous interpretations.

For the interpretative processes to begin, students need to perceptually structure the field into figure, which becomes sign, and ground. Normative interpretations can only emerge and be arrived at when the relevant signs have become figure such that their relation to the referent can be elaborated through one or more interpretants. Our own research shows that even scientists do not perceptually isolate those features that would lead them to a correct interpretation of graphs. Thus, when we are unfamiliar with a situation, be it a graph or some other inscription, we do not know which of the different ways of looking at it will lead us to a successful and useful interpretation. Some of our readers might assume that the objects and events depicted thus far in this chapter are so simple that there is no way that they could be perceived in some different way. Let us take a look at the interaction of twelfth-grade students with the software sometime toward the end of their first hour with it. The students in the episode have explored the microworld; they have oriented the arrows in different directions and have given them different lengths, following these actions by running the simulation prepared in this way. The teacher (Wolff-Michael Roth) approaches one group, possibly sensing that the students have been orienting the arrows and being concerned that they might not figure out what is going on. Such sense and concerns lead call teachers to interventions, which allow both finding out what students know and redirecting their investigations so that they can 'learn their lesson'.

In the present situation, the teacher orients the velocity and force arrows up and down, respectively, and then asked what they would see (speaking turn 01). From his (a trained physicist's) perspective, the situation corresponds to a ball thrown upward (velocity arrow up) and eventually moves down towards the ground, as a result of the gravitational force acting between ball and earth. His is a pedagogical move that reduces the complexity of the task because both velocity and force are in the up-down direction only and have no component in the left-right direction (i.e., $v_x = 0$; $F_x = 0$). In other words, this is as if there had not been an upward dimension of the velocity in figure 9.3. Much like beginning to drive on a parking lot, lower complexity in physics instruction is thought to make it easier to learn significant patterns. (The * indicates at which time students and teacher see the image to the right.)

01	Teacher:	What if you had that point up?
02		(3.5) ((Moves velocity arrow straight up.))
03		And this one would be pointed like this? ((Moves force arrow to center of object, turns it pointing straight down.)) *
04		((Teacher runs the simulation, resulting in the configuration depicted.)) *
05	Carl:	It goes straight down.
06	Tom:	Yeah, it goes downward.

After the teacher has set the initial conditions (turn 03–04), he runs the simulation. Both Carl and Tom describe the circular object as having moved 'straight down' or 'go[ne] downward'. In contrast to the teacher, who has set up the simulation and has known what would be seen—the object moving upward, the velocity arrow getting shorter and shorter, disappearing, then growing again in downward direction—the two students see the object go down. At this point in the interaction it is not clear whether they attend to the velocity vector. Although we do not know at this time whether the two students also attend to the initial conditions of the arrows as set up by the teacher, let us assume for the moment they do. In this case, they will have to associate the object that has 'go[ne] straight down' with a configuration that has one (the velocity) vector point upward and the other (force) vector point downward. They may also attend to the final state of the velocity vector, which pointed down. In this case, their interpretations will pertain to explaining a downward motion only, but not the initial upward motion that has not been salient to them. That is, although the structure of the initial condition contains an upward vector, their interpretation will not (initially) include and explain how it is related to the motion observed.

From the perspective of our semiotic framework (figure 9.2), students have not structured the field in a way that gives them that particular sign required for arriving at an interpretation that is taken as the norm in the discipline of physics. It is not that students have enacted some faulty reason. Rather, they have not *perceived* the initial upward motion nor attend to the fact that the trajectory trace includes several circles above the starting position. Pushed for an explanation, the students will have to make a connection between the interpretant of the trajectory as perceived ($sign_1$) and the interpretant of the vector configuration ($sign_2$). Although they might construct such a connection, though however reasonable it might be given the signs they constructed, it cannot be a correct interpretation from the perspective of physics. That is, it may not be the interpretative process, for which students draw on the traces their life experiences have left, as the problematic point in instruction. Rather, students may not perceive those signs relevant to learning what they are supposed to learn; that is, the nature has to be taken as equivocal or its presence taken as the problem in and for instruction.

Increasing the salience

There are different processes or conditions that might enhance the possibility that students themselves begin to question the nature of the sign. For example, another student present during the episode says, in the next turn, 'I think it went backwards first though'. That is, she observed the events differently, and articulates this difference in several ways within the same utterance. First, she uses the observational predicate 'went *backwards*', which contrasts Carl's 'goes *straight* down'. A second contrast exists in the adverb 'first', which contrasted the lack of an articulation of different parts of the event. A third difference was set up by the contrastive or adversative use of the adverb 'though'. Thus, following the two previous speakers, who provided the same description of the events, this new speaker articulated opposition, which opened up the possibility for different ways of viewing the trace ('sign') or at least for a debate in which alternate ways of seeing what is to be interpreted would be articulated. Here, differences in the nature of the sign become apparent in the descriptions of perceptions provided by different participations.

Having multiple participants present is not a sufficient condition for differences in apparent signs to become salient. For example, in one of our study, we observed four students watch objects rolling down an inclined plane. After the first time, one of them describes what he saw as 'constant motion' (interpretant). All four students in the group continue rolling the same and other objects, and note the similarity with the previous event. One student then formulates the hypothesis that objects roll down an inclined plane with constant velocity (interpretant of all previous interpretants). His peers agree. Ten minutes later, after having continued to roll various objects, he all of a sudden says, 'It is going faster', and immediately thereafter asks, 'What is going on here? What happened to my hypothesis?' He notes and his three peers agree that objects rolling on the inclined plane actually accelerate. That is, although there were different individuals involved, they all attended to the same perceptual trace (sign), although they later consider it to have been wrong.

Changing the rate at which events occur also allows different things to become salient, leading to different signs and interpretants. This was the case when the teacher in the previous section re-ran the simulation, but now in slow motion (speaking turns 01–03), or actually, by going step by step through the recorded trace.

01	Teacher:	Here we can play that slowly, slomo– look like this, Okay, this is where we started our three two one, do you see this? ((Brings the simulation back to its starting point, then plays one step at a time.
02		do you see what happened? This ((points to lower object image)) is where we started out. Look what happens to that small arrow ((Moves simulation one step further.))
03		What happens? Didn't you liken it to somebody to something– if you throw a ball up, what happens?
04	Carl:	It will come back down; it slows down, eventually.

The teacher repeatedly asks students to attend to changing display (turn 01) and in particular, to the small arrow (turn 02), and to compare the unfolding events to the starting point ('This is where we started out' [turn 02]). The teacher then asks students to recall what they have said between this and the earlier episode, that ball will come back down. Here, Carl, who initially responded to the question, attends to the object, describing its motion as 'slowing down'. He also adds the qualifying adverb 'eventually', which means that 'it slows down' has the sense of stopping. Here, then, the student sees the trace, which

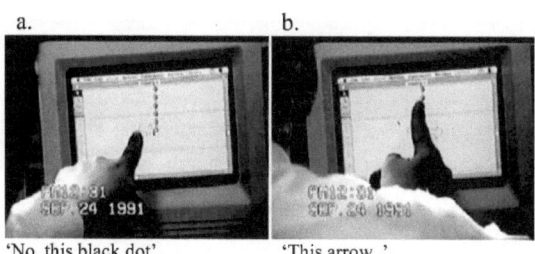

'No, this black dot'.   'This arrow..'.

Figure 9.4. The transcript and video illustrate how (a) indexical (pointing) and (b) iconic gestures—the finger resembles the arrow on the monitor—are used to make salient particular signs, which, in their relationship to interpretants, serve to elaborate the sign-referent relation.

unfolds in a step-by-step fashion, in a particular way; it becomes a changing sign that is elaborated by the interpretant 'slowing down'.

Such evidence makes it clear that we cannot take for granted that students actually receive information that some instructor intends to supply. If our observations are correct, then one of the important questions we need to answer is how students eventually break out of this seemingly hopeless situation. In moving towards an answer we suggest that the use of pointing (deictic gestures) and imagery gestures (iconic gestures) allows students to converge on a common perception of objects and events. In chapter 5 we articulate different ways in which teachers and lecturers make salient certain aspects of photographs using these gestures in a variety of ways, leading to different functions of gestures. In face-to-face conversations among students, too, deictic and iconic gestures play an important role of making sense *together*. Deictic gestures can be thought of as depicting lines that connect narrative point of view and targeted object. Prior to the episode in figure 9.4, another student suggests that grabbing it at the 'black dot on the end' would turn [force]. Now, Mike disagrees and places his index finger on the heel of the arrow (figure 9.4.a) (not visible on the video image is the black dot 'handle' that is normally visible to the user of *Interactive Physics™*). Here, then, pointing is used to disambiguate the referent of 'black dot [at the end]'.

Iconic gestures resemble their referents in some way; that is, they often provide a visual image of an object or thought. In figure 9.4.b, Mike holds up his right index finger and utters 'this arrow'. At this moment there are two arrows

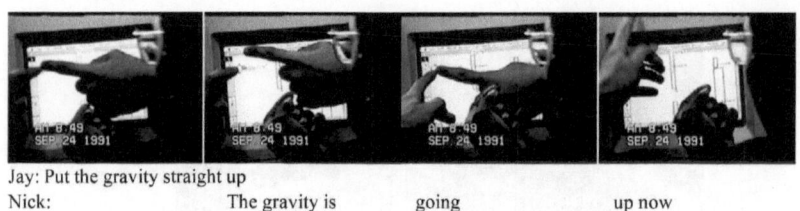

Jay: Put the gravity straight up
Nick: The gravity is   going   up now

Figure 9.5. The transcript and video illustrate how (a) indexical and (b) iconic gestures are used to make salient particular signs. Gestures are especially important when students have not yet developed a common language for articulating descriptions and explanations of objects and events.

on the screen: the force (open) arrow points to the right and slightly downward, whereas the velocity (small) arrow points straight up. When Mike says 'this arrow' the two listeners (as well as the analyst of the video image) are most likely to pick an arrow that somehow resembles the shape of the gesture. In figure 9.4.b, this is the velocity vector (from the perspective of the camera hidden behind the finger). Gestures that depict the motion of entities are also called iconic because the shape of the hand's trajectory has visual resemblance with the trajectory of the object. Thus, iconic gestures are used to denote objects and their movements across the monitor. Iconic gestures also depict conceptual entities (velocity, force) that are said to explain the events perceived. Here, there is a close connection between the actions (pointing, mouse manipulation) on objects and gestures, and the nature of the objects perceived.

Deictic and iconic gestures that pick out signs by touching them or picking out their shape in a perceptual way assist students in attuning each other to their respective experiences. Picking out objects and events and negotiating the signs of the descriptive and explanatory language are the starting point for the learning process in which students evolve a new language about moving objects.

The following episode allows readers to understand the salience and importance of deictic and iconic gestures (figure 9.5). Here, we see the hands of three students Edward, Jay, and Nick. Because of their initial difficulties understanding each other and making sense out of each other's talk, the three began to enact many gestures. These allow them to become attuned to the way they used particular words and the objects they were intended to refer to.

Jay, sitting to the far right, intends to suggest a new experiment. Because their changing ways of referring to the objects became confusing, he actually

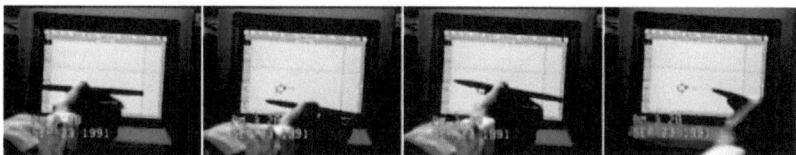

Like when we are doing it–   It goes in that straight direction.

Figure 9.6. The transcript and video show how the hand and pencil configuration and movement constitute interpretant signs that are related in iconic ways to an arrow and the trace of the object.

moves his finger straight up from the object. Nick, who begins to speak before Jay has finished, moves his index finger and hand upward from the force. Both situations involve iconic gestures. In the second case, there are two aspects of iconic nature. The hand (see last panel) is shaped in the arrow configuration and also moves upward. Potentially, the later aspect already encodes a hypothesis about the outcome of the investigation to be conducted in the proposed configuration. But more importantly, the hand is in the shape of a pointer configuration, in the shape of an arrow, and therefore constitutes an interpretant of the arrow on the screen. In this sign-interpretant relation, where similarity is based on iconicity, sign and interpretant each contribute to motivating the salience of the upward direction. That is, the hand shaped in the way it is, pointing upward, aids in pointing out an important aspect of the sign on the screen, and in this, aides in its perception or rather, in the perception of this aspect.

The ability to describe an event, that is, to make propositional statements that connect things and verbs arises after the ability to pick out simple objects from the environment and represent them with some arbitrary sign. Our analyses show time and again that verbal forms of description arise after gestures have been used to depict the focal events. In the episode depicted in figure 9.6, Mike attempts to describe what is happening when they run the experiment involving [force] and [velocity] lined up horizontally. The video images show that the verbal description of the object ('it') as moving in a 'straight direction' follows after the student has already gestured the trajectory once. The text coincides with the repetition of the gesture in which the right hand sweeps from left (where the object is located) to the right until the hand moves out of the picture. There are therefore two forms of delay. We describe these delays below.

The present episode also provides us with some evidence as to the relationship between the different layers. In figure 9.5, Jay and Nick are talking about

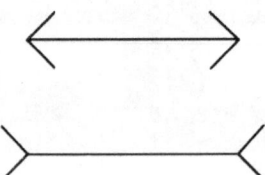

Figure 9.7. In this well-known illusion, the lower straight line is seen as longer, although measurement shows that it is of the same length as the line above it. Knowing of such effects allow us to compensate for the apparent illusion.

gravity and point to the arrow representing the force. They want it to point 'straight up' and see it as 'going up now'. At this point, we can think of their talk being about the force space as articulated in figure 9.3. There is nothing in the two utterances that allows us to assume a layering to have occurred. However, as soon as they set the simulation in motion, they are talking about the moving object and the velocity and force simultaneously, that is, they no longer separate the conceptual world used to model the phenomenal world. This is apparent in figure 9.6, where Mike's hand iconically depicts the movement of the object, while holding the pencil in the direction of the velocity arrow. Here, his hand itself constitutes a form of layering, phenomenal object, its mathematical position space, and the velocity space intimated in the pencil. The hand and pencil constitute an iconic sign that serve as interpretant of the inscription visually available on the computer monitor.

In the layering that occurs in such situations, the conceptual world of position, velocity, and force spaces are fused with the world of our experience. This world is always given to us in our perception in its factual nature. We do not usually interrogate it—unless we know that for some reason, our perceptions are being deceived, as this happens the case of known situations that lead to illusions (e.g., figure 9.7). That is, when Carl and Tom say that the object 'goes straight down', we have to take this as a description of how the world appears to them. It is on this perception that the two will base their reasoning. When the conceptual worlds are fused onto the phenomenal world, we then face the problem that we have to regard them as equally factual. That is, by fusing the conceptual worlds onto the phenomenal world, we make an epistemological commitment and see concepts as true rather than as constructions that allow us to navigate and explain the world of our experience. In this case, the (conceptual) maps become the territory that they represent.[3] But the two are different, and confusing them comes with the same dangers that come with mistaking play and situations that it pretends to portray. Shooting up an opponent in a videogame is one thing, shooting up your teacher and peers because you do not like them for one or the other reason is a whole different ball game. In the present day world, the difference between the two is no longer as distinct as we might want it to be—the shooting at Colombine High School or the recurrent killings of some employer by disgruntled (former) employees show just how close the worlds of videogames and the world of our experience have become.

The point we make here is that *critical* graphicacy involves being able to interrogate the conceptual worlds for what they are. This move is more easily made if students do not conflate phenomenal and conceptual worlds. The physics students in the Introduction, who discuss the nature of fields as phenomenal or conceptual entities are well on the way to make the distinction, and therefore on the way of developing a stance that allows them to critically interrogate conceptions more generally and inscriptions more specifically. That is, these students are well on the way of developing *critical* graphicacy, which allows them to question the politics involved in way inscriptions are deployed as part of ongoing activity.

The layers in praxis

In the present software, two different types of entities come to be layered and therefore risk to be conflated or, more importantly, risk that students come to believe in an isomorphic relation between the mathematical structure implicated in the vectors (arrows) and worldly objects and events, here embodied in the circular object. Not all software modeling Newtonian physics is written in this way. Jeremy Roschelle has developed the *Envisioning Machine*, a piece of software with a split screen, in which students can see objects and their movements in one window, and point masses to which force and velocity vectors were attached in the other window.[4] Thus, the *Envisioning Machine* clearly separates these ontologically different domains, one containing signs referring to experiential objects and events in the phenomenal world, the other containing signs denoting epistemic entities.

Returning to the present situation, *Interactive Physics™* layers the two domains, which therefore requires additional work to separate them in the learning process in the way we have seen the students in chapter 1 separating what can be seen, the iron filings in patterned arrangements around a magnet, and the magnetic fields that are used to explain these patterns. As critical educators interested in providing the resources for the development of a critical graphicacy rather than merely indoctrination into a particular graphicacy, we believe in aiding students to separate the two domains, because it is only at that point that they recognize that knowledge and concepts could be different. It is only when they recognize this possibility that they can search for alternative signs and in the possibility of negotiation and creation of alternative worlds. Is there evidence how the twelfth-grade students came to work with two layers? Let us return and observe one of them attempt to explain what is going on in this microworld.

At the moment of the episode, the three students still use different names for the arrows and use them in inconsistent ways. Students previously associated them with 'time', 'energy', 'time step', and many other words. In figure 9.8, then, Glen attempts to describe and explain the events (traces are still visible in the top left of the first frame). His utterances (figure 9.8) are paralleled by the gestures of both hands. Glen holds his right hand with fingers parallel to the open arrow [force]. He then makes another brief circular gesture, which marks the transition between two iconic gestures and highlights the salience of the hand, while uttering 'that arrow' that immediately preceded the causal meaning unit 'that's why it is pushing it. . .' Before he says 'the velocity' (third frame in figure 9.8.c) his left hand appears, held parallel to [velocity]. In the next frame, both hands are visible: the right parallel to [force], the left parallel to [velocity]. Then, the right hand 'pushes' against the left hand, which is moving to the left. This movement continues to the end of the sentence and out of the video frame. Here, the gesture of the right hand (figure 9.8.c) begins 0.83 seconds (i.e., 0.10 + 0.20 + 0.53) before the corresponding word 'pushing' (figure 9.8.f). That is, the iconic gesture provides a visual description of the object trajectory (still visible in figure 9.8.a) that Glen attempts to explain before he actually verbalizes the explanation. Though sophisticated for the moment in which Glenn articulates his understanding, the episode also shows a similar fusion of map and territory as the one we discussed in the previous section.

Much like Mike, Glen uses his hand as an iconic interpretant of the events on the monitor. However, in contrast to Mike, Glen models all layers at once using the two hands. His left hand, barely visible in figures 9.8.c–e, is initially pointed to the right and then, like the velocity arrow on the monitor, begins to rotate as hand and arm move to the left. The right hand is also in a pointer configuration (figures 9.7.b, d, e). Clearly visible is the fact that the hand keeps its spatial orientation, that is, always points in the same direction although the hand, too, moves to the left. This hand shows the same behavior recognizable in the force vector on the monitor. In addition, both hands describe a parabolic trajectory. That is, the two hands model the four different layers, the left hand standing in for the object, its mathematical position, and its velocity. The right hand

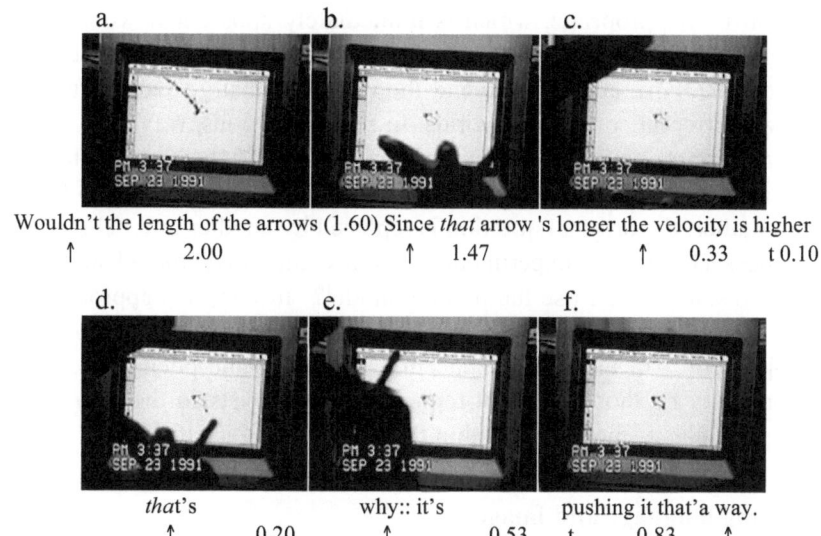

Wouldn't the length of the arrows (1.60) Since *that* arrow 's longer the velocity is higher
↑   2.00   ↑   1.47   ↑   0.33   t 0.10

that's   why:: it's   pushing it that'a way.
↑   0.20   ↑   0.53   t   0.83   ↑

Figure 9.8. Transcript and associated images from the videotapes. The following tran-scription conventions are used: ↑ = *marker that aligns text and image above it;* t = *time marker;* (1.60) = *1.60 seconds pause; italics for stress of syllable [e.g.,* that*'s];* :: = *lengthening of phoneme; and 0.53 = 0.53 seconds between markers.*

reproduces the mathematical position onto which the force representation has been layered. Not only are layered inscriptions seen as one, but also they have been enacted by the same person and therefore have become part of the same factual world given to the experiencing person. Force and velocity, though clearly belonging to conceptual worlds, have become fused with the world of experience. Whereas this layering might assist students in learning about New-tonian physics, not being able to distinguish the phenomenal and the conceptual worlds puts them at a disadvantage in that it becomes difficult to replace the lat-ter when necessary. Why, the reader may ask, might it be necessary to distin-guish between phenomenal and conceptual worlds? Let us take a look at another example from physics.

High school students learning about the nature of light are frequently con-fused because light behaves, under certain circumstances, like a wave; under dif-ferent circumstances, it behaves like a moving particle. Because instruction does not distinguish between phenomenon and its representation, students are con-fused as they consider the question, 'Is light a wave or is it a particle?' Evidently, the question is inappropriate, which is immediately apparent if we distinguish the phenomenal and the conceptual worlds. Light, brightness on the retina, can be experienced. *Wave* and *particle* and the drawings that go with them are aspects of two different conceptual worlds. In some situations, wave and associate concepts are appropriate for *explaining* what happens, whereas in other situations, the alternate set of concepts is appropriate. The question then no longer is the same. Rather, the original question is replaced by another one, 'Under which circumstances (for which experiment) do I use the wave model and 'Under which circumstances do I use the particle model?' In fact, this approach would provide an inroad to *critical* graphicacy, as students could engage in considering the appropriateness of alternative conceptions of the same phenomenon. These conceptions may be those held but replaced by physicists in the course of the history of the field, or those held within the student body itself.

Coordinating language and image

The development of an appropriate language for making observational and theoretical descriptions lags with respect to the gestural communicative forms. Thus, students are able to gesture the relationship between the instantaneous velocity of an object (denoted as small arrow on the monitor), force on the object (denoted as open arrow) and the trajectory that the object follows, one or two consecutive lessons prior to evolving an equivalent form of talk. Our videotapes show how the lag between gesture and language decreases to the order of seconds and eventually disappears; gestures and corresponding speech then occur simultaneously (within ± 200 milliseconds).

At the time of the episode in figure 9.8, Glen (along with his two peers) does not yet describe the arrows in scientific terms, that is, as force and velocity. As our evidence recorded two weeks later shows, he and his two peers use the appropriate scientific (verbal) language only two weeks later during a subsequent lesson with the microworld. However, although he has not yet developed an appropriate language at the time of figure 9.8, his gesture is already consistent with scientific representation practice—when understood as a description of the relationship between the concepts of velocity and force. He characterizes the action of the outline arrow as 'pushing', which is a vernacular form of describing forces. Glen also associates the longer pushing arrow with a resulting higher velocity. Here, the referent of 'velocity' is not completely clear and two readings are possible. Because the utterance coincides with the positioning of the left hand, 'velocity' can be heard as the referent to the left hand: therefore, the longer right arrow (force) pushes more and therefore leads to a longer left arrow (velocity). But the fragment 'Since that arrow 's longer the velocity is higher' can also be heard in the sense that the longer right arrow is equivalent to a higher velocity. Here, then, 'velocity' (incorrectly so from a scientific perspective) would refer to the right arrow. However, the referents for each of the two hands are clear by their position in space in the course of the motion. The directional orientation of the right hand is constant and parallel to [force]. The left hand changes direction in the way [velocity] previously changed.

Tool for making conversation and constructing interpretants

The present chapter shows that *Interactive Physics™* can facilitate and structure in important ways students' conversations and learning in science classrooms. The presence of the computer provides students with opportunities to achieve topical cohesion, clarify the meaning of ambiguous terms, and focus each others' attention to pertinent aspects of the display. Classical studies in conversational analysis show that topics of talk are maintained in part when participants directly respond to each other in consecutive utterances. However, it is well known that conversations constructed mainly by the temporal proximity of each utterance tend to drift. The computer display affords a possibility for constructing a coherent conversation. Because it is often difficult to make sure that all participants talk about the same thing as they converse about conceptual issues in real time, the physical presence of the object of talk, here the inscription, provides students with a means for coordinating just what they are talking about. The immediacy of the displayed inscription allows pointing to specific objects and the animation affords a replay of motion by means of gestures. Through this immediacy, the interface becomes an object of conversation immediately available. Thus, the conversation is about something actually present rather than about some object removed from the students' experience. Students coordinate their conversation directly over and about the display, which can be said to constitute an anchor for the conversational topic. They test their understanding and repair mishearing in the presence of the object of the conversation.

It is important to note that the computer microworld is not just another inscription teachers can use in their classroom. Here, the computer microworld is used as a way to support the interactions in a group and to facilitate the negotiation of understanding not simply as another means of showing students information they are to learn. Because students see real world events and inscriptions in different ways than scientists and science teachers, the latter need to use such tools as the computer microworld to facilitate interactions rather than as carrier of unambiguous information that students are to take in.

The ability to point to and repeat in gesture specific events are important aspects of inscriptions that facilitate the communication between people, and more specifically, scientists. In this respect, inscriptions (computer displays, drawings, electron-micrographs, read-outs from various measurement instruments, or engineering designs) are crucial elements in the sense-making and communication activities in science and technology.[5] These devices are so important to the discourse in science and technology that engineers will interrupt meetings to get design drawings or get devices that would allow them to render the design in sufficient accuracy before they will continue with their meeting; or scientists will try to at least render a drawing in gesture in order to be able to make sense of their respective contributions to the on-going theory talk. If these devices were to exclude some individuals, the communication would be severely interrupted.

Students' interactions in the context of *Interactive Physics*™ share many similarities with those involving a different kind of inscription, which is also designed to provide them with opportunities for talking science: concept mapping. Concept maps and computer displays have a temporally and geographically local character: (a) they bring students together to work on some common goal, although in many cases this common goal or product is as yet to be worked out in its detail (and is worked out through the interaction); (b) they engage participants in their own construction, and do so increasingly because there is more to be pointed to, talked about, and encoded their own historical development; and (c) they represent 'common' history for the participants, and common ground. Thus, *Interactive Physics*™ and concept mapping can be conceptualized as inscription-based tools for social thinking. They serve as a means of organizing the task, the task environment, the discourse in which students engage to construct the design, and the final understandings. This organization is achieved, because the displays are part of a *taken-as-shared* problem space in a double sense. First, this problem space permits students to work simultaneously on the same task. Second, the displays also provide for a *taken-as-shared* conceptual space in which the participants can refer to common objects by means of words, drawings, or gestures. The participants' individual involvement is indicated by their attention to the objects and the other participants of this *taken-as-shared* space.

Indoctrination and the reflexive stance

The computer facilitates students' coordination of events (moving objects) with the changing vector diagrams in important ways. In traditional teaching situations (as on the blackboard), students get to see only a very small number of diagrams for a specific event. A problem students are often asked to answer in research on their conceptions, a penny is thrown up into the air, and students are asked to explain forces on a penny at three places: halfway up, at the top of its flight path, and half-way down. Science educators assume that students easily envision the transformation of one situation into another, that is, that students visualize the transformation of the velocity vector and the constancy of the force vector between the three positions. Science educators also assume that students (non-experts) map the conceptual framework (vector diagrams) onto the empirical world. However, ample conceptual change research has shown that students are far from mapping natural events in terms of a Newtonian framework that can be expressed by vector diagrams. The microcomputer world provides an inscription that associates vector diagrams directly with events, thus making arrangements for the coordination of ontologically different entities (three conceptual and one phenomenal). However, whereas the microworld allows students to co-ordinate representations of conceptual and phenomenal objects, science educators must not assume that students proficient in microworld activities will now map the inscriptions onto the empirical world (as this might appear in chapter 8), students have to engage in a difficult dual-reverse transformation. They have to transform the representations of forces and velocities (vectors) into the realm of the conceptual and the circular microworld object into the phenomenal world. In this dual transformation students easily lose the correspondence between the two elements that was established in the microworld by visible connections. It is only through the implication of the real world into the conversation about the microworld that students may be able to maintain this correspondence. Going back and forth between real world experiments and the computer microworld to model the observed phenomena appears to be a plausible teaching strategy to facilitate students' learning. In addition, such a back and forth would constitute an interesting background for engaging students in discussions of epistemological nature concerning the work of science as one of modeling and representing of real world phenomena.

At the time we conducted the research on learning in the context of *Interactive Physics*™ initially, we thought that the software aided students in the learning of Newtonian physics. In fact, it assisted students' sense-making efforts as the physical presence of the simulation supported various kinds of conversational acts and it permitted students to coordinate the phenomenal and the conceptual. To understand how the microcomputer world achieves this, it is helpful to compare it to the two different origins that have led to its display design, static vector diagrams and empirical experiments.

These results, while interesting from the classical science educators' perspective, are, in some respects, highly problematic from our perspective as critical educators interested in developing not just graphicacy, the competency to read and use inscriptions, but *critical* graphicacy, the competency to construct and deconstruct the politics of inscriptions in the construction of the world. On epistemological grounds we hasten to add some caveats to this view of the microworld. The computer display, which layers two inscriptions denoting entities that are different in their nature lead students to form the idea that physics knowledge constitutes an eidetic image (from the Greek *eidos*, shape) that is not merely a mental picture, but taken as the transcendental essence, the truth of an object-in-experience. As such, the computer display would appear to show in some obvious and self-same manner the way nature really is because it hides the constructive work by means of which the phenomenal and the conceptual are coordinated. In the same way, the link that maps microworld onto real world events is an achievement and not an ontological *a priori* condition. Teachers who adhere to a constructivist epistemology may have to spend some additional time to discuss with students the constructed nature of the various correspondences discussed here.

10   Epilogue: steps toward *critical* graphicacy

Throughout this book, we present analyses (a) of the work of reading required for understanding scientific inscriptions in different domains and (b) of students actually engaged in reading such inscriptions. However, our aim as critical educators is not just the provision of opportunities for students to become graphically literate; rather, we want students to develop *critical* graphicacy, that is, we want them to become literate in constructing and deconstructing inscriptions, the deployment of which is always inherently political. In this epilogue, we elaborate some of the steps we might want to take in the project of assisting students to develop not just graphicacy but *critical* graphicacy.

We begin by providing a brief case study from our research that exhibits what we have in mind when writing about *critical* graphicacy. We then argue that educators ought to take an anthropological perspective on learning, which allows them to conceive of *critical* graphicacy as practice, inherently public and shared, rather than as stuff in individual minds. We provide an example from a physics course taught by one of us, which shows what critical (science) educators might want to do in their own classrooms to develop *critical* graphicacy. We close with recommendations for the redesign of inscriptions and for issues critical educators might want to consider in their own teaching.

What kind of *critical* graphicacy?

Readers might ask us about the kind of *critical* graphicacy we have in mind, and, perhaps, about what it looks like in concrete practice. As part of our research on science in the community, we have in fact observed what could be at least the beginning of a vision of what *critical* graphicacy might look like: People in the community interrogating the inscriptions that they are confronted with in and over contentious issues that affect their lives. In the following, we relate a brief excerpt from a public meeting in the community of Central Saanich during which local residents questioned the decision by their town council and mayor

241

not to build a water main that would connect their homes to the water grid that supplies all other homes in the community. The citizens do not succumb to the scientists' attempt to domineer the public meeting with 'science speak' and the associated scientific inscriptions resulting from their research, but critically interrogate a number of these inscriptions, articulating their shortcomings and the fallibility of interpretations. This questioning of scientific inscriptions in public debates constitutes one aspect of the kind of *critical* graphicacy that we have in mind.

In Central Saanich, water issues are and have been a central and ongoing concern. In this particular case, the media had repeatedly reported on the situation in one part of Central Saanich, Senanus Drive, which is not connected to the water main. All properties of Senanus Drive supply their own water from wells on their properties and from cisterns. Because the wells are recharged mainly through precipitation, the water supply depends on weather patterns; very dry summers lead to depletion of the aquifer and a correlated increase in the mineral contents and contamination by biological organisms. Repeatedly over the past decade, local newspapers have carried stories about the fact that the water quality in these wells had been declared unfit for consumption without prior boiling, forcing residents to drive five kilometers to get their water from the next gas station.

The town council of Central Saanich felt that the estimated cost of $850,000 (in 1999) for extending the currently existing water lines to supply Senanus Drive was too high to be covered through its allocated and available budget. A total of six reports had been commissioned prior to the public hearing. These reports included one by the Regional Health Authority, a preliminary report by the Central Saanich Water Advisory Task Force, the report by a consulting hydrologist, the final report by the Water Advisory Task Force, a minority report submitted by a subgroup of the Water Advisory Task Force, and a report by the municipality. The mayor of Central Saanich had called for a public hearing, including the authors of the technical and scientific reports. These authors would first provide a sketch of their work and subsequently make themselves available to respond to questions and comments from the public. Furthermore, the hearing was to provide opportunities for members of the community to ask questions and to make presentations.

One of the technical reports was produced by a local engineering firm Lowen Consulting. During the hearing, Dan Lowen had made his presentation by drawing on the rhetorical registers that characterize science and engineering: It was very factual, presented those aspects of the methodology that supported claims about the generalizability of the results from the nine sampled wells to all

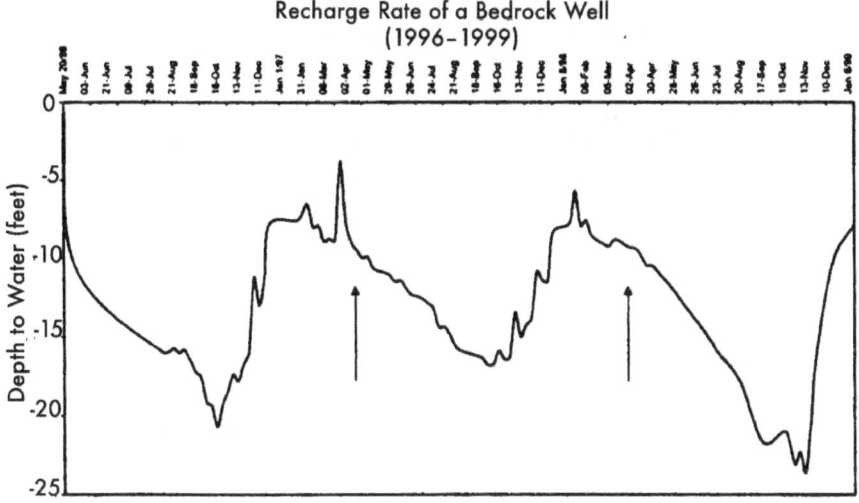

Figure 10.1. This graph represents the depth to the water level in one local well. The scientific expert in a public hearing had sampled the water quality in April. At issue in the present episode is the extent to which the April values can be taken as representative of water quality over the year. The arrows point to the April values of depth to the water level in two consecutive years prior to the expert's sampling.

wells at Senanus Drive. He described the point of sampling to support claims that in fact well water had been tested rather than water that had remained in pipes, storage tanks, or cisterns. The privileged status of this expert witness was established by presenting him as 'a professional engineer and a professional geologist', including him into the ranks of other scientific experts, some of whom were introduced with their degrees. ('Mr. Yin has a Masters of Science degree, and has significant experience with water quality issues and he has been involved extensively in both reports in the sampling episodes'.) This, of course, is a classical tactic for constructing exceptional status of the claims made by the individuals thus elevated.

In the following excerpt, a local resident, Knott, asks the expert Lowen to evaluate his own results in the light of those apparently contradictory ones presented by other scientists in the service of the regional health authority. The particular issue in the questioning is Lowen's claim that the well water sampled in April constitutes something like an average value in terms of the amount, and therefore, in the concentration of chemical elements. Featured in both exhibition and report was a graph displaying the water levels in a local well over a three-year period (figure 10.1). Here, the major issue is whether Lowen's data represent an average value or whether they have to be interpreted as a short-term, best-case scenario. Knott not only asks Lowen to make this evaluation but also, as his further questioning shows, brings out the pertinent issues that have led to the contradictions between the report authored by Lowen, on the one hand, and that of the scientists from the regional health authority, on the other.

01	Knott:	How do you feel about your results now that you've heard the person from the Ministry of Health describe what he feels could affect the readings. You seem to rely very heavily on the readings that you took. Okay, which seems to be explained by the difference in rainfall. Now are you in agreement, for example, that the aquifer and the water coming from the wells is largely the result of rainfall?
02	Lowen:	That's right. It's all the result of rainfall, not just largely.
03	Knott:	Okay, well there's a buffering effect, there's an immediate effect. I mean, well I shouldn't say that there's an immediate affect, a slower affect, one that occurs as a result of a longer period of time. We'll say, predominantly two, three to five months?
04	Lowen:	All the water that got, all the fresh water that got in the aquifer came from rainwater, originally.
05	Knott:	Yeah, but–
06	Lowen:	Some may have been in the ground longer than others– a longer period of time and water that has been in the ground for a longer period of time can have a higher mineralization because it absorbs minerals from the rocks.
07	Knott:	Of course. And what we have– Can you tell me the years that you have charted here, what years were those logged for?
08	Lowen:	Yeah. The, the observation well has been in service since, nineteen seventy-nine but I can see here, and that's from October ninety-seven and we had data up to June of nineteen ninety-nine.
09	Knott:	Okay, so what is your understanding of what happened last fall and this spring with respect to the water amounts of rainfall? Was this a heavy period or a normal period?
10	Lowen:	I know that it was a record period per rainfall but it's not reflected in water levels in the area because the peak water levels in the aquifer in, nineteen eighty-eight were higher than in the winter levels in nineteen ninety-nine.
11	Knott:	Just a minute. You just said that it was a direct result of water and we've just had a record rainfall and it doesn't affect it? Well, there's something missing here.
12	Lowen:	That means that only a certain amount of the rainfall can get into the aquifer being the heavy rains are running off. That's my interpretation.

13 Knott: Well, it could also–
14 Lowen: There's a limiting factor as to how much can get down into the–
15 Knott: Well, it's okay, this is true but the thing is that what we've experienced rainfall in the order of five hundred and twenty-two percent on average, as far as monthly averages are concerned, increase over the summer months. In other words what we've got through the winter period, through the five months previously preceding your test results. If you took that and compared that to an average summer month, a month through that period, there were five hundred and twenty-two percent more. Now, it would seem to me that we're probably not dealing on an average result with your tests, we're probably dealing on the hydrostatic head feeding that aquifer up in the higher, very much higher ends, so that the readings that you're getting are very much diluted.
16 Lowen: The hydrograph that we have shows that the water levels are average in late April, early May and I put the average water level on the hydrograph here and the–
17 Knott: Could it be an error? Could you be in error here?

It is evident from figure 10.1 that the groundwater level fluctuates throughout the year. Lowen, however, has measured the concentration of chemical and biological contaminants at one moment. Lowen previously has argued that his data—taken at a water level in the aquifer midway between its minimum and maximum values during the sampling year—represent an average and therefore representative value of the biological and chemical parameters of water quality. The scientists from the regional health authority suggested, on the other hand, that there are fluctuations in water levels such that during one half of the year, the quality values are in fact below the Health Canada standards. From this position, at issue is not whether Lowen's data represent the average concentration but the fact that during certain periods of the year the concentration is above the Canadian standards. This is the same period when the residents are advised not to consume their water; this therefore deeply affects their quality of life. In this excerpt, Knott questions Lowen about the variations in the graph, that is, the level of water in the aquifer at the time Lowen conducted his measurements. In this, this interaction constitutes a moment where the contradictory claims of acceptability of water quality are made salient again in public—at issue is the interpretation of a graph.

Critically interrogating inscriptions such as graphs is exactly what we have in mind. Knott already exhibits the kind of *critical* graphicacy that we might want students to develop. He is but another resident of the area but he asks Lowen to reflect on the results of his own readings after having another report, which has come to a different conclusion. In particular, Knott asks Lowen to attend to and interpret the effect of sampling moment to the amount of rainfall at and prior to the time of testing. In response to Lowen's description that all the water in the wells and in the aquifer comes from rainfall, Knott states that there should be a buffering aspect. That is, changes in the aquifer do not directly correlate with the rainfall but are delayed by three to five months.

As Lowen makes another categorical statement that all water comes from rain, Knott's interjection 'but–' makes Lowen retract or at least modify his earlier statement. He now admits that the concentration of dissolved mineral in water would increase if it stayed for longer amounts of time in the ground. Knott's subsequent question seems to aim at eliciting from Lowen a statement about the groundwater levels; Lowen has to admit that there have been record water levels but attempts to argue that these rainfalls do not affect the groundwater levels. However, Knott questions this claim by contrasting it with a previous, seemingly contradictory one. Lowen attempts to argue that there are limits to absorption and that much of the rainwater would be carried away as run-off—therefore not visible on the graph (figure 10.1). Knott is unsatisfied by this response. He suggests that there is a 522 percent increase in rainfall from the summer to the winter months (turn 15). The water levels could not have been average as Lowen claims in his report and that therefore, the concentrations of substances would generally be lower (more diluted) than under normal circumstances. Lowen, however, responds that the hydrograph is showing an average reading. We notice in the last turn that Knott is unsatisfied by this answer. He first casts doubt on Lowen's conclusion by evoking the possibility that an error could have been committed in using the average water level registered by the hydrograph in the months of April and May. More so, in a clever or astute rhetorical move, he squarely attributes the responsibility to the expert, 'Could you be in error here?'

In this episode, we see a member of the general public question the content of a scientific report on water quality, the groundwater-level graph, the methodology for gathering the data, and for the readings made on the basis of the data. Knott and all the other individuals who ask questions do not appear to be intimidated by the social status usually attributed to scientific experts (here, several scientists were introduced by mentioning degree and rank in their respective institutional hierarchies). They pursue their lines of questioning which put into relief what otherwise are presented as authoritative statements and inscriptions about the quality of the water they are using.

In this instance, Lowen uses scientific inscriptions as if they are inherently apolitical. This may be the case when he constructs and ponders them in his office. However, as soon as these inscriptions leave his office, for example, as part of his report to the town council of Central Saanich, they are political and are used for political purposes. Despite the plight of the Senanus Drive residents, recurrently depicted in various local media, the town council of Central Saanich is unbending. It does not provide resources or permission for the extension of the water main, though it gives such permissions for development elsewhere in the community. Its apparent motive is to curb the potential for changing the zoning by-laws in effect for Senanus Drive. At the moment of the debate, the zoning is rural. Once the water main was extended to the area, fire hydrants would also be installed, which then would allow a change in the zoning and the subdivision of the lots, and a build-up of the area. Fearing such a development, successive town councils in the past voted against the water main extension, despite changing zoning in other parts of the community from agricultural-land-reserve status to low-density housing. Here, the political nature of the scientific inscriptions becomes quite clear. The report Lowen has filed, including many scientific inscriptions, becomes a cornerstone in the decision-making processes the outcome of which continues to be the blockage of a water-main extension.

The political nature of the inscriptions also and especially became apparent in the public meeting, where they become points of contestation. In this arena, another important point becomes apparent: *critical* graphicacy is better studied and thought about as practice, something that real people actually *do*, rather than as something that is inaccessibly stored somewhere in their minds. That is, we encourage critical educators to take an anthropological perspective on *critical* graphicacy rather than the psychological and cognitive perspectives that are prevalent in government policies and reform documents.

Critical graphicacy: taking an anthropological perspective

In today's schools, students spend a lot of time on language-oriented literacy but they spend little time developing graphicacy, let alone critical graphicacy. However, we live in a visual world and inscriptions constitute a common feature in scientific texts and a core practice in the sciences. The difficulties students have in competently producing and reading graphs, in the standard ways envisioned by psychologically oriented educators and reform documents have been well documented. The approach that some of this research has taken so far is to relegate the problems to software and hardware of individual students' minds. Researchers write about retardation in cognitive development, disabilities, and cognitive deficiencies. A psychological approach will therefore make suggestions for teaching that focus on remediation, acceleration of development, or improvement of metacognition.

An anthropological approach takes a decidedly different tack. Researchers question attributions to the hardware of individual brains, pointing time and again to the socially situated nature of the practices in which most students have few opportunities to participate. When children and students are taught reading, their reading competencies are cultivated to bring them into the community of readers. Children are taught to reenact in their own reading the communal practices of reading; they become sentinels of society's way of reading. In the trajectory from initial reading instruction to competent reading, reading leaps beyond the textual basis to the thing that is transparently read. In contrast, few students are competent in transparently reading inscriptions. Our analyses throughout this book show that the frequency and quality of inscriptions used in textbooks do little to contribute to students' enculturation in competent practice reading graphs; they do even less to support the development of *critical* graphicacy, useful in deconstructing and recognizing the political nature of inscriptions. This lack may well be one of the sources of undeveloped competence in reading inscriptions. Our anthropological approach therefore refocuses investigators to ask questions such as, To what degree do younger and older students enact *critical* graphicacy?, To what degree do children deploy inscriptions to make arguments?, and To what degree does the learning environment allow students to develop their own goals and purposes that require inscription practices? Our advice to critical educators is therefore concerned with establishing learning environments where inscriptions are part and parcel of the everyday practices in which students engage.

From our anthropological perspective, some of the problems identified by cognitive-oriented research are artifacts of investigation. For example, students are said to take graphs as pictures. Such a use has much in common with the inappropriate use of words students learn from dictionaries without opportunities to use the words in practice. Our own research shows that even scientists inappropriately use iconic features as resources for answering questions about unfamiliar graphs. Iconicity is based on perception, and perception, as we show in chapter 9, gives us a world in all its concreteness and reality.

Words and sentences are not islands of meaning; nor are graphs. When displayed as part of some text, resources of various types, such as those found in captions and main text, assist in reading an inscription. It is ironic that students are asked to engage with graphs without such resources (e.g., chapters 2, 4, 7, 8). Scientists, on the other hand, are provided with these resources, as a matter of course, in scientific publications that constitute part of their daily work. It appears self-evident that the resources provided for reading an inscription, if they are to be of help, must be familiar to the reader. Yet research on inscription uses labels that are unfamiliar to students; for example, despite ample research that suggest the non-canonical use of 'velocity' by high school and college students, science education researchers use it to label axes. Within our framework, increased non-standard readings of graphs come to no surprise given the 'misconceptions' associated with the notion of 'velocity'.

An anthropological perspective on inscriptions suggests to teachers that they should not expect students to be competent after a unit on graphing, but that graphicacy should be an integral part of learning to do science. That is, students should be enculturated into graphicacy as an aspect of scientific forms of life, which includes reading, interpreting, and producing inscriptions. In the endeavor of enculturating students both to graphicacy and *critical* graphicacy, the tools they use and the activities they engage in should not be substitutes but allow students to engage in continuous trajectories toward competent participation in *critical* graphicacy. An analogy from language learning may clarify our point: Expecting that students read inscriptions consistent with some canon is as likely as expecting them to understand and use an unusual word, such as 'ontology', from a definition. Students need to participate in appropriate and recurrent conversations and reading activities to become familiar with the uses of the word 'ontology' in specific situations. In this book, we present evidence that students learn to use inscriptions when they need inscriptions for rhetorical purposes to convince their peers about the results of their own investigations (chapter 3), much in the same way that scientists do in their daily work.

Our book raises the question about possible developmental trajectories of *critical* graphicacy. On the surface, the inscriptions that appear in textbooks have, at some level, a family resemblance to those of books and periodicals written for sophisticated lay readers (e.g., *Scientific American* or *New York Times*). One might therefore expect that textbooks support the development of graphicacy different from those of scientific communities. However, our recent research shows that even many scientists have difficulties interpreting graphs typical for high school and university textbooks. Furthermore, many undergraduate ecology students and preservice science teachers indicate that they do not attend to or attempt to interpret graphs that they encounter in their readings. At this point, the pedagogical value of graphs in textbooks still remains to be established.

Our anthropological perspective makes important suggestions to researchers concerned with students' competencies in reading, interpreting, and making graphs. For example, some researchers miscue students in their attempts to read graphs. This practice lies on a polar opposite to the normal practices in science where many redundancies are built into the presentation of graphs to decrease the associated ontological ambiguity. As we point out in chapter 2, scientists produce graphs with captions and main text such that they minimize other interpretations. Chapter 4 shows that students' interpretations of photographs changes with the addition of textual resources, both caption and main text. In scientific publications, where there is a chance that readers are not from the identical domain, enough resources (instructions, descriptions) are provided so that readers recover but one interpretation, that intended by the authors. We acknowledge that scientists develop the competence to recover the intended interpretations; they do so because of the shared background assumptions and extensive experience with specific phenomena that they bring to the task. Our perspective therefore encourages science education researchers to make sure their own inscriptions are complete and supply the necessary resources to make sense although students may not share their background assumptions.

With respect to our semiotic analyses, we suggest to assist students in developing critical graphicacy to conduct such chemisemiotic analyses. These could begin with simple questions, 'Hey, what is this diagram all about?' which they, with some assistance by their teacher, can then attempt to answer. They could then move on to ask critical questions such as, 'Is this diagram attempting to make us look at the phenomenon in a particular way?', 'Are there other ways of portraying the same issues that would bring out different aspects?', and 'What are the epistemology and politics involved in this diagram?' In the process, they will learn some chemistry, which they can use for their benefit as they employ it in further critical analysis. For us, the point of schooling at the junior and high school levels cannot be indoctrination to this or that discipline, with indoctrination in the attendant disciplinary blind spots and ideologies. Rather, what citizens of tomorrow's world need are knowledgeability that allow them to deal *critically* with different forms of knowledge, in part embodied in inscriptions, which require critical graphicacy to be interrogated. At this point, critical educators might ask us, 'What can I do to assist me in teaching for critical graphicacy? and 'How might I go about teaching critical graphicacy?'

Helping students develop a more reflective stance

Developing a more reflective and even critical stance toward inscription means doing epistemology. From our perspective, epistemology is the most needed and yet underrepresented subject in all of formal education. Yet how are students to become critical, how are students to develop competent critical graphicacy if they never have opportunities to participate in epistemological praxis. We might ask, 'Are students indeed able to do epistemology and deconstruct inscriptions, scientific or otherwise?' We not only respond with a resounding 'Yes!' but also, in this section, provide some brief excerpts from a physics course taught by one of the authors (Wolff-Michael Roth). This course featured as an integral component discussions pertaining to epistemological issues, intended to assist students in taking a critical stance toward knowledge and the different ways in which it is represented. The purpose was not for students to develop a particular epistemological position—objectivism, realism, social constructivism, radical constructivism, or any other ism we can think of—but rather to develop the discursive tools to take a reflexive stance toward whatever position they want to take.

Epistemology is usually presented as a body of abstract knowledge, out of reach for most people. Accordingly, the essence of epistemology is thought to be attainable only by those who have special capabilities. In this view, epistemologists as philosophers more generally are the high priests of bodies of sacred knowledge shrouded in the languages of the past. But, we might ask, 'What might happen if we created opportunities for everyone to interpret the sacred books and inscriptions they use in describing and explaining the world?' 'What might happen if we take away the cloak of special and prerequisite capabilities?' or 'What might happen if we question the nature of epistemology as a (sacred) body?' Once we begin answering these questions, we begin to realize that we will be left with epistemology as praxis, that is, a form of lived experience witnessable by participants and observers (researchers, analysts) alike. Thus, epistemology can be documented and analyzed rather than being some ethereal object located in sacred books and in the minds of high priests. In other words once we consider epistemology as an everyday praxis, accessible to and open to be enacted by human beings in general, epistemology looses its air of divine mystery. These issues become apparent in the following episode.

The twelfth-grade physics students in this episode have been reading a variety of texts concerned with the invention of physics language and, therefore, the invention of reality. Here, they discuss with their teacher the relationship between what experiments allow us to see and the theories that explain the experimental outcomes. The concrete case they discuss is the apparent dual nature of light, and the interaction participants make reference to a diagram on the chalkboard (figure 10.2). Matt, who opens the episode, makes the distinction between an inaccessible aspect of the world and the theories and concepts used to describe and theorize it.

01 Matt: Laws such as the one for the double slit ((points toward the chalkboard [figure 10.2])) aren't infinite and the complexity, it's just the simplification of the nature to it, the level at which our puny little brains can under-

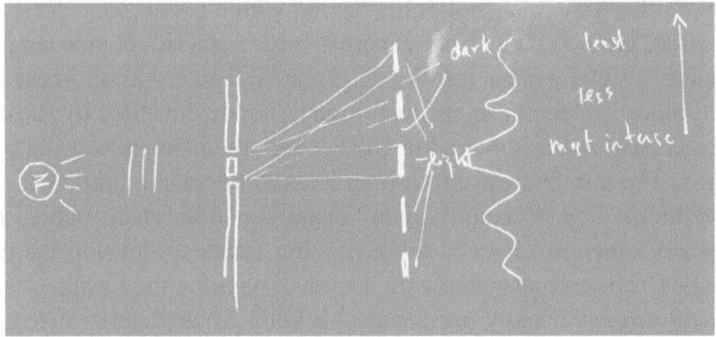

Figure 10.2. The students and teacher pointed to this chalkboard inscription as part of their discussion of the nature of light and electrons, and the relationship of diagrams and language to the two phenomena.

		stand. It's not real, we are just applying a conventional viewpoint, wave or particle, to totally unconventional subjects, like light and electrons, and then we describe them in our limited terms. And that is not the way they deal with–
02	Teacher:	That is where we use our limited terms and images.
03	John:	We use our experience to describe what we see.
04	Teacher:	How is that when we use our limited experience?
05	John:	A paradox. When you take something like that you can't just simply, something that is a wave and a particle at the same time, which shows the same characteristics, like electrons behave like the light ((points to the chalkboard)), and both needn't be necessary, both are equally valid and both indispensable, so we, above and beyond our experience to describe...
06	Nick:	our experience...
07	John:	what is happening, whereas it is not a wave *and* a particle.

The teacher then comments that what needs revision are statements that attribute definite wave and particle images to phenomena. In this statement, the equation of wave and particle images and the phenomena, constructed in the particle 'is' (which the teacher stresses in his utterance), requires a reflexive investigation; in fact, this equation ('*that*') requires a revision. Here, the teacher provides opportunities for and invites students to question the reification of knowledge and the world.

08	Teacher:	So the question really is– In your first sentence you said that something *is* a wave and something *is* a particle and *that*'s really what needs revision, isn't it?
09	Nick:	It can't be either...
10	Matt:	it can be described by...
11	Nick:	but it actually isn't either of them.
12	Todd:	Yea.
13	Teacher:	Do we know what it is?
14	Matt:	It's like when you are talking about electrons. I like the example [Gregory] gives us, and what you drew on the board ((points to the chalkboard)): you have a marble with a screen and the hole gets smaller then knowing its exact position won't tell you where the electron is even though you know it's path better. With the electron it's the opposite, that's just like your first reaction is, 'Gee that is weird. Why would that happen? That doesn't make sense'. But it makes sense according to what– Why should our limited experience be the standard by which all concepts are judged? Like the real path that electrons have to have is not like marbles, that's just like that's just like a–
15	Marc:	How can we interpret what we don't know? Like we don't have no basis to know?
16	Hun-Jun:	Well, that's what I was saying!
17	Marc:	We can't picture it.
18	Nick:	We can try to.
19	Todd:	Yea.
20	Matt:	It's like position and motion aren't characteristics of electrons this–
21	Teacher:	That picturing doesn't feature here remember! These pictures ((points to the chalkboard)) get us off the track.
22	Nick:	I can't picture it.

Taking turns, Nick and Matt articulate the fact that phenomena such as light and electrons can be described—not articulated here under which conditions this would be the case—in terms of waves and particulates but that the phenomena are not identical their description (turn 11). The teacher insists: if the phenomena are not identical to their descriptions—which is the way in which scientific knowledge usually is represented, that is, as depicted in the isomorphism {fundamental structure ⟷ mathematical structure}—what then are phenomena. Matt begins a longer explanation about viewing electrons as particles and having to face the conflicting result that narrowing their passage way through a wall (as represented in the chalkboard inscription [figure 10.2]) actually leads to less information about their pathway than more. The narrower the hole is through

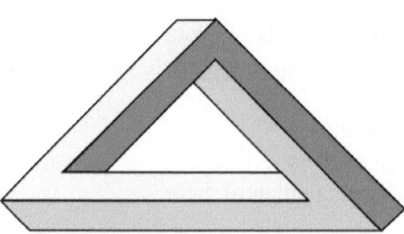

Figure 10.3. Many students and their teacher are familiar with the artist M. C. Escher, and his many drawings featuring impossible objects. The impossible existence of real three-dimensional objects arises from the reduction from three dimensions of real objects to the two dimensions of this inscription.

which the electron is made to pass, the more scattered the places where it would actually be found if a photographic plate was positioned behind the hole.

Marc then asks the really pertinent question from an epistemological perspective: if we do not know, how can we interpret (or know) what we do not know? What can we know if we pull the rug from under the assumption that there is a definite relation between the world and our pictures and language of its true nature? Matt constructs another definite attribute, now its absence: position and motion *are not* characteristics of electrons. But the teacher reminds him and his peer that the picturing needs to be questioned: the picturing gets them off the track (turn 21).

The teacher makes a move to link the discussion to art, but Matt preempts him naming the artist and articulating the fundamental issues dealt by him in his drawings—illusions and different dimensions of space. Different students then contribute to the conversation, evoking particular drawings by the artist, such as the impossible triangle similar to the one we represent in figure 10.3, and which the teacher sketched on the chalkboard during a different lesson involving the same students. Considering what happens when a three-dimension object is represented in an inscription of lesser dimensions, for example, two, allows the students to understand and explain the distortions that result from representing complex phenomena in a small number of dimensions.

23   Teacher: There is an artist—
24   Matt: Three- and four-dimensional space— I have that book over there in my apartment where some of his pictures are shown, may be you know Escher, M. C. Escher, who shows seeming illusions, namely buildings that twist around itself.

25	James:	Is that in the triangle?
26	Matt:	The stairs that go upwards to end where they begin.
27	Todd:	There is a waterfall.
28	Teacher:	And the waterfall goes back to the start. Now why would that be?
29	Chris:	It's the shadings that create the illusions.
30	Teacher:	But what happens, what does he represent?
31	James:	Three-dimensional space in two dimensions.
32	Teacher:	Three-dimensional space in two dimensions, as soon as you try to simplify a complex phenomenon–
35	Matt:	You get it distorted.
36	Teacher:	And something very similar happens in the quantum world. See there is a common phenomenon: we try to use our way of describing and what arises is something that looks like a ball in some cases and in other cases it appears as if it was a wave. But we have to say, we just don't know. We cannot—I think that's what seems to come out of this book ((*Inventing Reality*))—we cannot know what the world really is and so all we can do is give descriptions, which always use language and images.

After naming the contents of some of the Escher drawings and paintings (turns 24–27), the teacher asks students about the origins of these illusions. Chris suggests that the shadings create the illusion (e.g., figure 10.3), which is a technical answer about creating drawings and paintings on a flat piece of paper that seem to portray three dimensions. The next speaker, James, makes the epistemologically interesting opening to the statement, completed by Matt, that the illusions are the result that three dimensions are represented in two. The teacher then summarizes and extends the students discussion of the art into physics. Because we only have our representations, we cannot know what the world really is. Quantum phenomena appear like balls in some cases and like waves in others. Unlike with the Escher drawings, where we know real triangles, stairs, and waterfalls that we can relate to the way they are inscribed, we have no access to the quantum world other than through the way it is inscribed.

This episode provides a glimpse of the kind of classrooms we envision for the development of *critical* graphicacy. Here, the teacher is not merely concerned with allowing students to use and read physics inscriptions, but he puts students into a situation where they begin to interrogate inscriptions in a critical way. This is evidenced throughout the physics course, where students, for example, are asked to make decisions about choosing one of several mathematical equations that equally well fit the data they had collected. Having to make such decisions students come to realize that inscriptions are inherently political, favor one theory over another, introducing individual preferences, thereby undermining the claimed 'objective' nature of scientific inscriptions. This is apparent in the following statement produced by one of the sixteen-year old physics students:

Science is a language game. It allows us to talk about the world in community of knowers, which shares a common language. This language allows us to create tools—concepts and theories—to talk about this world, predict and explain events, and thus create our knowledge of this world. We are now forced to ask ourselves, 'What shapes do these tools and truths take and how are they used by us?' The answer takes us to the beginning of one of my essays where I stated that it is 'with words, with sounds, all joyful, playful and obscene' on which scientific knowledge is based. The language we create and use to describe our observations becomes the tool itself. By changing the language we not only change the law and principles science is stating but we also change a previously accepted truth and effectively make a new one. Thus, it is language and the way in which we choose to define the phenomena we observe that is at the core of our knowledge; it is through these words that we arrive at the images and ideas that allow us to predict and explain our observations. This holds true for everything in our lives, it is through our language that we communicate our ideas, thoughts, and feelings, and it is also through them that we are able to learn through the recreation of our perceptions within our minds.

After two years in the physics courses, where students engage in many discussions about knowledge and the politics of representation, they display sophisticated levels of *critical* literacy and graphicacy. They begin to recognize, as one student said, that 'science textbooks are like the bible—they are based on and require faith'. This, we suggest, takes us a fair bit toward the kind of citizen we envision, a citizen who can and do interrogate the practices of companies like Monsanto attempting to sell us on genetically modified organisms and other products that are not sustainable over long periods of time. We envision a citizen who is able and willing to engage scientists critically, which requires graphicacy in addition to a critical stance.

What might educators and textbook authors want to do to support an education toward such a vision? In the following section, we sketch some possible ways of how to go about supporting students.

Teaching with inscriptions

Our analyses of the way in which inscriptions are used in the textbooks have some implications for (a) textbook authors and (b) textbook readers, including teachers and students. Our work implies that textbook authors and publishers attend to the appropriate integration of the different inscriptions so that they in fact assist students in making sense. Further, the use of single inscriptions, for example, often does not allow a reader to disclose what really matters; (time) series of or contrasting inscriptions, on the other hand, make salient variation that are more likely to lead readers to identify the crucial and learning-enhancing aspects. Some important topics to which textbook authors should pay careful attention when selecting inscriptions are briefly addressed here.

Every inscription should have an appropriate caption associated to it via an indexical reference in the main text. An appropriate caption may contain more than just an identification of the object or phenomenon presented in the inscription and referred to in the main text. The caption should add enough text such as to guide the reader through a perceptual analysis that lies at the basis of any interpretation of the inscription, identifying relevant details and associating the figure with the main text. The figure (inscription and caption) should be explicitly associated with the main text via an indexical reference, preferably placed just after the first time the object or phenomenon of interest is mentioned in the main text. This would allow the readers to easily relate figure and text, significantly narrowing the possibilities of misinterpretations of the inscription. Time-series or sequences of photographs are preferable to single inscriptions to represent complex phenomena or changes. Pairs of inscriptions that allow comparison or that provide a magnified view of an especial detail of an object (or a broader view of a small object) seem to work better than single inscriptions.

When making use of a single inscription, special attention to the inscription itself is necessary. The background of the inscription should be easily recognized as such, especially when we deal with photographs and naturalistic drawings; the object or phenomenon of interest represented in the inscription should be highlighted against its surroundings. Centering and focusing the relevant object or phenomenon are ways of emphasizing what is important. Making use of neutral backgrounds, for example black backgrounds, is an efficient way of highlighting the object or phenomenon when it does not suffer from decontextualization. In some situations, arrows and other signs directly placed over the inscription can also help to distinguish important details.

Pertaining to readers, anecdotal information shows that while perusing textbooks students and teachers often attend little to inscriptions, especially photographs and their captions. If, however, there is information to be obtained from the dialectically related caption and inscription, students miss out on an important resource for understanding the topic that they study. Teachers and students need to pay more attention to the possibilities of inscriptions to enhance understanding of the accompanying texts. Perhaps the inscriptions could become themselves objects of discussions in the way students discuss other inscription in the previous section, which would allow students and the teacher to develop insights about how and what others see when they look at an inscription, and how they interpret it in the context of various other textual sources provided on the page. Further study, however, is needed to address these and other important issues related to the pedagogical potential of the inscription for students and teachers who make use of textbooks. We suggest that future studies may focus on students' and teachers' interpretation of inscriptions (cum caption and main text) in real time.

When students are reading textbooks, the way in which they interpret inscriptions greatly influences their understanding of the concept presented by the textbook. Thus, teachers should pay more attention to the complexities and subtleties of the processes involved in the interpretation of inscriptions when visual aids are used during their classes or on the textbooks they adopt. As it we make salient throughout this book, inscriptions can be confusing; captions and main texts may lack information necessary for students to connect inscriptions and text; students' previous knowledge and experiences influence their reading of the text and figures. Therefore, teachers need to be aware that students not only need to develop subject matter literacy, but also graphicacy to fully understand their textbooks.

New perspectives should be able not merely to explain known problems, but should be fruitful and provide descriptions for better approaches. We therefore provide (a) an alternative for the line graph about natural selection (figure 2.4) and (b) suggestions for teaching and researching graphing practices.

We redesigned the previously analyzed line graph on natural selection and the associated caption and main text (figure 10.4). This redesign takes as basic assumption that, ontologically, scientific concepts are heterogeneous assemblies consisting of the totality of re-presentations in words and inscriptions. Knowing a scientific concept therefore means being able to use (read, interpret, produce) its various aspects. For example, provided the distributions in figure 2.4 are accepted and shared inscriptions in the context of natural selection, knowing 'stabilizing selection' and 'directional selection' means being able to competently read *these* line graphs in *this* context. Our suggestion therefore includes the resources necessary to read normal distributions in the context of natural selection (figure 10.4).

An inspection of figure 10.4 shows that it includes many of the resources outlined in our ontology (see chapter 2): there are (a) labels, scales, and units on both axes; (b) markers, identifiers, labels to lines and points, and arrows; and (c) additional titles ('Environment A', 'Environment B') that set the plates apart. The caption provides explicit instructions for reading this graph so that the ex-

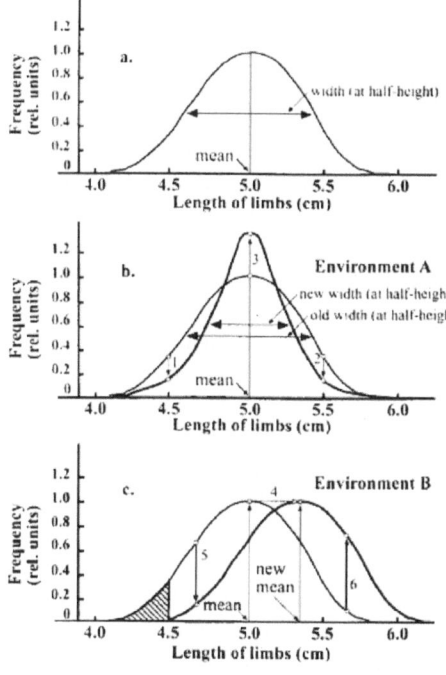

Figure x. Changes in the distribution of limb length in a hypothetical adult animal population over many generations. A. The original populations was characterized by a mean length of 5.0 cm limb length. b. After some time in Environment A, a new distribution is observed (bold line). The mean limb length is still 5.0 cm. However, there are fewer animals with shorter limbs (s) or with longer limbs (2), but more animals with a limb length near the mean (3). The width at half-height of the new distribution has decreased. The concept 'stabilizing selection' captures this shift from the original to thee new distribution. c. After some time in Environment B, a different kind of change is observed. The new distribution has the same shape and size, but its mean has shifted to about 5.3 cm (4). Notice that the frequency of animals with short limbs has decreased (5), but the frequency of longer-limbed animals has increased (6). Adult animals with limbs shorter than 4.5 cm, which existed in the original population (their total number is proportional to the cross-hatched area), over many generations are almost never observed under the conditions in Environment B. Thus, a 'directional selection' has occurred favoring longer-limbed individuals.

Figure 10.4. Suggested replacement for the graphs, caption, and main text featured in figure 2.4. Caption and main text are designed to provide a maximum number of resources to assist readers in understanding graphs and two forms of natural selection.

planation of stabilizing selection and directional selection are congruent with a reading of the graph. For example, the caption makes explicit how in Environment A the narrowing of the distribution is achieved when there are fewer individuals with limb length far from the mean. This reading then becomes congruent with a verbal description of 'stabilizing selection' in the main text that emphasizes favorable conditions for individual adults with limb length around the mean, and less favorable conditions for individuals far from the mean.

MAIN TEXT: Observed changes in the original distribution (figure x.a) of a genetically mediated aspect in a population, such as limb length, are modeled by means of the concepts of environmental conditions and selection. When the observed changes in the distributions lead to a narrower and higher curve (figure x.b), one can build the following model. Environmental conditions are favorable to the survival of animals with limb length around the mean (~ 5.0 cm), but less favorable to animals with longer (e.g., 5.5 cm) or shorter (e.g., 4.5 cm) limb length. Consequently, the frequency of animals with limb length around 5.0 cm increases, whereas the frequency of animals with much longer (e.g., above 5.5 cm) or much shorter (e.g., below 4.5 cm) limbs decreases. We can describe this phenomenon as 'stabilizing selection' to emphasize the increase in frequency of those animals with a characteristic around the mean. One can also observe changes in the distribution where the mean length of limbs changes, e.g., increases as depicted in our model ecology of the population in Environment B (figure x.c). When we observe such a shift in the distribution of adult animals, we model this as follows. The increase in the frequency of animals with longer limbs, with a concurrent decrease of animals with shorter limbs is explained by assuming that long-limbed animals have a better chance of survival to the age of reproduction in Environment B than shorter-limbed animals leading to the greater numbers that we observe in graphs like figure x.c. Seen from the perspective of the original population, the selection can be described as 'directional'. Shifts in the distribution of characteristic aspects of animals such as that displayed in figure x.c can be observed during times when the environment is changing rapidly (relative to the average age of the animal).

Our caption and main text also render explicit that the graph re-presents observations on hypothetical animal populations in different environments. In the main text, we use the verbs 'describe' and 'model' to make the theoretical function of the terms 'stabilizing selection' and 'directional selection' explicit. In this way, our Cartesian graphs allow readers to participate in graphing practices in a new way: Because of the tight integration of graphing and discourse practices, there is a great potential that they become resources for one another, stabilize each other, and therefore are more easily appropriated. Also, because our graph incorporates resources in the Cartesian graph, caption, and main text in a way similar to scientific journals, students are provided with opportunities to participate in authentic scientific representation practices. Our suggestion therefore provides for a stepping-stone on the trajectory to those inscription practices in which scientists engage on a routine basis.

The differences in length and detail between the original figures and text and our suggested replacement immediately stand out. At this point, figure 10.4 constitutes a hypothesis about a 'better' or 'more meaningful' graph that supports students' interpretations given their currently available reading practices. Empirical studies comparing students' interpretation practices remains to be conducted. A corollary of our suggestion would be a necessary reduction in the number of concepts that could be treated by textbook authors if the length of textbooks remained constant. The frequently dictionary-like treatment of scientific terminology would retreat in favor of more integrated discourse and inscription practices that are common in scientific communities.

Notes

Preface

1. Lori L. DiGisi and John B. Willett, 'What high school biology teachers say about their textbook use: A descriptive study', *Journal of Research in Science Teaching,* Vol. 32, No. 1 (1995), 123–142.

Introduction

1. John Berger, *Ways of Seeing* (London: Penguin Books, 1972), at 10.
2. Bruno Latour, *Science in Action: How to Follow Scientists and Engineers through Society* (Milton Keynes: Open University Press, 1987).
3. Frances K. Aldrich and Linda Sheppard, 'Graphicacy: The fourth "R"?', *Primary Science Review*, Vol. 64, No. 1 (2000), 8–11.
4. Pierre Bourdieu, 'The practice of reflexive sociology (The Paris workshop)', in Pierre Bourdieu and Loïc J. D. Wacquant, *An Invitation to Reflexive Sociology* (Chicago: University of Chicago Press, 1992), 216–260.
5. Greg Myers, 'Every picture tells a story: Illustrations in E. O. Wilson's *Sociobiology*', in Michael Lynch and Steve Woolgar (eds), *Representation in scientific practice* (Cambridge, MA: MIT Press, 1990), 231–265, at 244.
6. Michael Lynch, 'Laboratory space and the technological complex: An investigation of topical contextures', in S. Leigh Star (ed), *Ecologies of Knowledge: Work and Politics in Science and Technology* (Albany, NY: State University of New York Press 1995), 226–256, at 255, emphasis in the original.
7. See, for example, Richard E. Mayer and Richard B. Anderson, 'The instructive animation: Helping students build connection words and pictures in multimedia learning', *Journal of Educational Psychology*, Vol. 84, No. 4 (1992), 444–452.
8. Wolfgang Schnotz, E. Picard and A. Hron, 'How do successful and unsuccessful learners use texts and graphics?', *Learning and Instruction*, Vol. 3, No. 3 (1993), 181–199.
9. Latour op. cit. note 2, at 258.

10. Ken Morrison, 'Some researchable recurrences in discipline-specific inquiry', In D. T. Helm, W. T. Anderson, A. J. Meehan, & Anne W. Rawls (eds), *The Interactional Order: New Directions in the Study of Social Order* (New York: Irvington, 1989), 141–158, at 145.
11. Morrison, op. cit. note 10, at 147.

Chapter 1

1. Michel Foucault, *Surveiller et punir: Naissance de la prison* (Paris: Gallimard, 1975).
2. An interesting collection of such studies were published in the collection *Representing as Scientific Practice*. Michael Lynch and Steve Woolgar (eds), *Representing as Scientific Practice* (Cambridge, MA: MIT Press, 1990).
3. Henry A. Giroux, *Border Crossings: Cultural Workers and the Politics of Education* (New York: Routledge, 1992), at 244.
4. See, for example, Jacques Bertin, *Semiology of Graphics: Diagrams, Networks, Maps* (Madison: University of Wisconsin Press, 1983); Fernande St. Martin, *Semiotics of Visual Language* (Bloomington: Indiana University Press, 1990; and Edward R. Tufte, *The Visual Display of Quantitative Information* (Cheshire, CT: Graphics Press, 1983).
5. See, for example, Gaea Leinhardt and coworkers (1990). There are now researchers, including Valerie Walkerdine, who approach mathematical knowing as social practice, thereby de-centering attention from the grey matter 'between the ears and underneath the skull' to the individual's interactions with the social and material world. Gaea Leinhardt, Orit Zaslavsky and Mary K. Stein, 'Functions, graphs, and graphing: Tasks, learning, and teaching', *Review of Educational Research*, Vol. 60, No. 1 (1990), 1–64; Valerie Walkerdine, *The Mastery of Reason* (London: Routledge, 1988).
6. Howard Wainer, 'Understanding graphs and tables', *Educational Researcher*, Vol. 21, No. 1 (1992), 14–23, at 18.
7. Bruno Latour, *Science in Action: How to Follow Scientists and Engineers through Society* (Milton Keynes: Open University Press, 1987), at 258.
8. Umberto Eco defines a *sign* as a segmentation of matter pointing to and expressing something other than itself. Signs include letters, words, texts, pictures, drawings, and graphs. For the literary Sherlock Holmes, even the positioning of some objects, a piece of hair on a sofa or the barking of a dog at night were signs referring him to some object, event, or phenomenon. In the mathematics education literature, the notion of 'symbol' is often used in the same sense as 'sign' in the semiotics literature. Umberto Eco, *Semiotics and the Philosophy of Language* (Bloomington: Indiana University Press, 1984).
9. Bruce Wake, 'Best bang for education buck debated at school board', *Peninsula News Review*, 11 December 1998, at 2.

10. As outlined below, in the perspective taken here, signs and texts are mutually constitutive to the phenomena they signify. Reality therefore does not lie *behind* the text but squarely in the *context*, that is, what goes *with* (Lat. *co-, con-, com-*) and is required by the text. See, for example, Richard Rorty, *Contingency, Irony, and Solidarity* (Cambridge: Cambridge University Press, 1989).

11. Willard Van Orman Quine, *From Stimulus to Science* (Cambridge, MA: Harvard University Press, 1995), at 48.

12. Jon Barwise, 'On the circumstantial relation between meaning and content', In Umberto Eco, Marco Santambrogio and Patrizia Violi (eds), *Meaning and Mental Representations* (Bloomington: Indiana University Press, 1988), 23–39.

13. Semiotic research in general takes the existence of the sign for the individual as unproblematic. Valerie Walkerdine is one of those who make the distinction by noting that for two people an expression such as 'no more' may be the same signifier (segmentation of matter) but a different sign. Valerie Walkerdine, 'Redefining the subject in situated cognition theory', in David Kirshner and James A. Whitson (eds), *Situated Cognition* (Mahwah, NJ: Lawrence Erlbaum Associates, 1997), 57–70.

14. Reader might be interested in reflecting on the deep epistemological implications arising from the contrast of Karen's statement, 'This is a clogged pipe' as a descriptor (interpretant) of a wiggle, and the caption 'This is not a pipe' in René Magritte's painting of a pipe. This painting was the topic of a famous essay of the same name by Michel Foucault. Michel Foucault, *This is not a Pipe*, trans. and ed. J. Harkness (Berkeley: University of California Press, 1983).

15. Eric Livingston, *An Anthropology of Reading* (Bloomington: Indiana University Press, 1995).

16. Wolff-Michael Roth and G. Michael Bowen, 'Complexities of graphical representations during lectures: A phenomenological approach', *Learning and Instruction*, Vol. 9, No. 3 (1999), 235–255.

17. Studies on the interpretation of graphs do not ascertain whether students are familiar with the phenomena nor, for these matters with translating between graph and phenomena, or with the conventions regulating the use of particular graphical sign elements.

18. Wolff-Michael Roth, Michelle K. McGinn and G. Michael Bowen, 'How prepared are preservice teachers to teach scientific inquiry? Levels of performance in scientific representation practices', *Journal of Science Teacher Education*, Vol. 9, No. 1 (1998), 25–48.

19. Such examples can be found, for example, in John Clement, 'The concept of variation and misconceptions in Cartesian graphing', *Focus on Learning Problems in Mathematics*, Vol. 11, No. 2 (1989), 77–87.

20. See, for example, Paul Ricœur, *From Text to Action: Essays in Hermeneutics, II* trans. K. Blaney and J. B. Thompson (Evanston, IL: Northwestern University Press, 1991).

21. See, for example, Winfried Nöth, *Handbook of Semiotics* (Bloomington: Indiana University Press, 1990).

22. Because any phenomenon is 'captured' in one sign form or another, because all of our reflective knowledge requires representations, signs therefore only point to other signs. Constructivists come to the same conclusion: because we do not have access to the world as such but only to our descriptions of it, one sign can only refer to another. In a post-modern reading, we can therefore never escape con/text.

23. Barwise, op. cit. note 12.

24. Jenny Preece and Claude Janvier, 'A study of the interpretation of trends in multiple curve graphs of ecological situations', *School Science and Mathematics*, Vol. 92, No. 6 (1992), 299–306.

25. Giroux, op. cit. note 3, at 243.

26. Elinor Ochs, *Culture and Language Development: Language Acquisition and Language Socialization in a Samoan Village* (Cambridge: Cambridge University Press, 1988), at 16.

Chapter 2

1. Wolff-Michael Roth, *Toward an Anthropology of Graphing: Semiotic and Anthropological Perspectives* (Dordrecht, The Netherlands: Kluwer Academic Publishers, 2003).

2. Bertin provides an extensive overview of those aspects most frequently considered. See Jacques Bertin, *Semiology of Graphics: Diagrams, Networks, Maps* (Madison: University of Wisconsin Press, 1983).

3. Greg Myers, 'Every picture tells a story: Illustrations in E. O. Wilson's *Sociobiology*', in Michael Lynch and Steve Woolgar (eds), *Representation in Scientific Practice* (Cambridge, MA: MIT Press, 1990), 231–265, at 238.

4. Françoise Bastide, 'The iconography of scientific texts: principles of analysis', in Michael Lynch and Steve Woolgar (eds), *Representation in Scientific Practice* (Cambridge, MA: MIT Press, 1990), 187–229.

5. Wolff-Michael Roth, 'Emergence of graphing practices in scientific research', *Journal of Cognition and Culture,* (in press).

6. Bastide, op. cit. note 4, at 208.

7. Eric Livingston, *An Anthropology of Reading* (Bloomington: Indiana University Press, 1995), at 55.

8. Robin S. Reid and James E. Ellis, 'Impacts of pastoralists on woodlands in South Turkana, Kenya: Livestock mediated tree recruitment', *Ecological Applications*, Vol. 5 (1995), 978–992, at 987.

9. D. E. Moody, 'Evolution and the textbook structure of biology', *Science Education*, No. 80, Vol. 4 (1996), 395–418.

10. Roth, op. cit. note 1.

11. Sylvia S. Mader, *Inquiry into Life* 4th ed. (Dubuque, IA: Wm. C. Brown, 1985), at 504.

12. Mader, op. cit. note 11, at 504.

13. Raymond F. Oram, *Biology: Living systems* 4th ed. (Columbus, OH: Charles E. Merril, 1983), at 649.

14. Oram, op. cit. note 13, at 82.

15. Oram, op. cit. note 13, at 649.

16. See G. Michael Bowen, Wolff-Michael Roth and Michelle K. McGinn, 'Interpretations of graphs by university biology students and practicing scientists: towards a social practice view of scientific representation practices', *Journal of Research in Science Teaching*, Vol. 36, No. 9 (1999), 1020–1043.

17. William A. Andrews, B. Jennifer Andrews, D. A. Balconi and N. J. Purcell, *Discovering Biological Science* (Scarborough, Ontario: Prentice-Hall Canada, 1983), at 568–570.

18. Andrews et al., op. cit. note 17, at 568–570.

19. See, for example, Wolff-Michael Roth, G. Michael Bowen and Domenico Masciotra, 'From thing to sign and "natural object": Toward a genetic phenomenology of graph interpretation', *Science, Technology, & Human Values*, Vol. 27, No. 4 (2002), 327–356.

20. Jay L. Lemke, 'Multiplying meaning: Visual and verbal semiotics in scientific text', in J. R. Martin and R. Veel (eds), *Reading Science* (London: Routledge, 1998), 87–113.

21. T. Piersma, J. van Gils and P. de Goeij, 'Holling's functional response model as a tool to link the food-finding mechanism of a probing shorebird with its spatial distribution', *The Journal of Animal Ecology, 64* (1995), 493–504, at 497.

Chapter 3

1. See, for example, Jean Lave, *Cognition in Practice: Mind, Mathematics and Culture in Everyday Life* (Cambridge, England: Cambridge University Press, 1988).

2. Wolff-Michael Roth, Kenneth Tobin and Kenneth Shaw 'Cascades of inscriptions and the re-presentation of nature: How numbers, tables, graphs, and money come to re-present a rolling ball', *International Journal of Science Education*, Vol. 19, No. 10 (1997), 1075–1091.

3. In *Descartes' Error*, Antonio Damasio provides evidence from neurological studies that individuals with particular forms of brain damage continue to do well on intelligence tests, tests involving considering alternative actions, and tests of social skills, but utterly fail in making decisions in everyday situations. That is, knowledge displayed in testing situation is not the same as practical knowledgeability. See Antonio Damasio, *Descartes' Error: Emotion, Reason, and the Human Brain* (New York: HarperCollins, 2000).

4. G. Michael Bowen designed and taught the unit together with Wolff-Michael Roth, who conducted the research in the classroom, designed and administered tests, videotaped lessons, conducted interviews, transcribed recordings, and analyzed the data.

Chapter 4

1. An ecotone is defined as an ecological community of mixed vegetation created by the overlap of adjoining communities; it is usually a belt rather than an abrupt line. For example, the floodplains of the Amazon River are sometimes interpreted as ecotones between upland and rivers, and sometimes they are viewed as specific ecosystems. But there is still a lot of controversy over the ecotone concept, in part because boundaries cannot be easily delineated.
2. Bruno Latour, *Pandora's Hope: Essays on the Reality of Science Studies* (Cambridge, MA: Harvard University Press, 1999).
3. John Law and Michael Lynch, 'Lists, field guides, and the descriptive organization of seeing: Birdwatching as an exemplary observational activity', in Michael Lynch and Steve Woolgar (eds), *Representation in Scientific Practice* (Cambridge, MA: MIT Press, 1990), 267–299.
4. In Roth, Bowen, and McGinn (1999) and Pozzer and Roth (2003), we provide detailed analyses of the different types of inscriptions used in high school textbooks and, in the former article, compare it to the types of inscription found in scientific journals. See also chapter 2. Lilian L. Pozzer and Wolff-Michael Roth, 'Prevalence, function, and structure of photographs in high school biology textbooks', *Journal of Research in Science Teaching*, Vol. 40, No. 9 (2002), 1089–1114; Wolff-Michael Roth, G. Michael Bowen and Michelle K. McGinn, 'Differences in graph-related practices between high school biology textbooks and scientific ecology journals', *Journal of Research in Science Teaching*, Vol. 36, No. 9 (1999), 977–1019.
5. The frequencies of these four functions among a total of 148 inscriptions were: decorative ($n = 8$ [5.4 percent]), illustrative ($n = 52$ [35.1 percent]), explanatory ($n = 42$ [28.4 percent]), and complementary functions ($n = 46$ [31.1 percent]).
6. George Lakoff provides examples of a category that are nearer the center, and therefore more prototypical for a category versus those that are nearer the peripheries of two categories, and therefore more problematic in their assignment to one or the other. See George Lakoff, *Women, Fire, and Dangerous Things: What Categories Reveal about the Mind* (Chicago: University of Chicago Press, 1987).
7. Multimedia such as web pages allow different ways of accessing images, for example by making available a 'button' linked to an image so that the reader can, if desired, make the image appear in a new window, which itself may be moved around the monitor.

Chapter 5

1. Reviews of this literature in anthropology, linguistics, and education are provided in Wolff-Michael Roth, 'Gestures: Their role in teaching and learning', *Review of Educational Research*, Vol. 71, No. 4 (2002), 365–392; Wolff-Michael Roth, 'Gesture-

speech phenomena, learning and development', *Educational Psychologist*, Vol. 38, No. 4 (2003), 249–263.

2. Jay L. Lemke, 'Multiplying meaning: Visual and verbal semiotics in scientific text', in J. R. Martin and R. Veel (eds), *Reading Science* (London: Routledge, 1998), 87–113.

3. Charles Goodwin, 'Professional vision', *American Anthropologist*, Vol. 96, No. 4 (1994), 606–633.

4. Wolff-Michael Roth, G. Michael Bowen and Michelle K. McGinn, 'Differences in graph-related practices between high school biology textbooks and scientific ecology journals', *Journal of Research in Science Teaching*, Vol. 36, No. 9 (1999), 977–1019.

Chapter 6

1. John Berger, *Ways of Seeing* (London: Penguin Books, 1972), at 10.

2. See, for example, Lucy A. Suchman and Brigitte Jordan, 'Interactional troubles in face-to-face survey interviews', *Journal of the American Statistical Association*, Vol. 85 (1990), 232–244; Wolff-Michael Roth and Yew Jin Lee, 'Interpreting unfamiliar graphs: A generative, activity-theoretic model', *Educational Studies in Mathematics* (in press).

3. On of the studies in our research team showed how environmentalists used the same photograph to show (a) how a particular creek had been modified and devastated by human practices and (b) how some citizens in the community are already taking care and monitoring the creek. See Stuart Lee and Wolff-Michael Roth, 'How ditch and drain become a healthy creek: Representations, translations and agency during the re/design of a watershed', *Social Studies of Science*, Vol. 31, No. 3 (2001), 315–356.

4. For examples of such studies see Schoultz, Säljö and Wyndhamn or Ueno and Arimoto. One of the studies in our research group was explicitly designed to investigate the different ways in which graphicacy was contextualized, both in the experience of the participating physicists, on which they drew to elaborate potential referents, and the interview itself, to which the participants were oriented, and which mediated the responses they provided. See Roth and Lee, op. cit. note 2; Jan Schoultz, Roger Säljö and Jan Wyndhamn, 'Heavenly talk: Discourse, artifacts, and children's understanding of elementary astronomy', *Human Development*, Vol. 44, No. 1 (2001), 103–118; Naoki Ueno and N. Arimoto, 'Learning physics by expanding the metacontext of phenomena', *The Quarterly Newsletter of the Laboratory of Comparative Human Cognition*, Vol. 15, No. 1 (1993), 53–63.

Chapter 7

1. See, for example, Wolff-Michael Roth, Carolyn Woszczyna and Gilian Smith, 'Affordances and constraints of computers in science education', *Journal of Research in Science Teaching*, Vol. 33, No. 10 (1996), 995–1017.
2. The reader in this study refers to the interpreter of a given inscription. We analyze the interpreting work required to get the point of an inscription, e.g., Boyle's law for figure 7.1, in a perspective of the researcher or the author, but by making us unfamiliar with the inscription.
3. Semantics is the interpretation or meaning of a sentence, word, etc. (The Canadian Oxford Dictionary, 2001). In this chapter, we refer to semantic model as a model that is used to analyze the work of interpreting an inscription.
4. Wolff-Michael Roth and Reinders Duit, 'Emergence, flexibility, and stabilization of language in a physics classroom', *Journal for Research in Science Teaching*, Vol. 40, No. 9 (2003), 869–897.
5. Jacques Derrida, *Limited Inc* (Chicago: University of Chicago Press, 1988).

Chapter 8

1. See, for example, Yehudit J. Dori and Mira Hameiri, 'Multidimensional analysis system for quantitative chemistry problems: Symbol, macro, micro, and process aspects', *Journal of Research in Science Teaching,* Vol. 40, No. 3 (2003), 278–302.
2. See, for example, Dorothy L. Gabel, 'Research on problem solving: Chemistry', in Dorothy L. Gabel (ed), *Handbook of Research on Science Teaching and Learning* (New York: Macmillan Publishing Company), 301–326.
3. Shimshon Novick and Joseph Nussbaum, 'Junior high school pupils' understanding of the particulate nature of mater: An interview study', *Science Education*, Vol. 62, No. 3 (1978), 273–281.

Chapter 9

1. Michael Lynch, 'Method: measurement—ordinary and scientific measurement as ethnomethodological phenomena', in Graham Button (ed), *Ethnomethodology and the Human Sciences* (Cambridge: Cambridge University Press, 1991), 77–108.
2. Paul Ricœur, *From Text to Action: Essays in Hermeneutics, II,* trans. K. Blaney and J. B. Thompson (Evanston, IL: Northwestern University Press, 1991).
3. Gregory Bateson, *Steps to an Ecology of Mind* (New York: Ballantine, 1972).
4. Jeremy Roschelle, 'Learning by collaborating: Convergent conceptual change', *The Journal of the Learning Sciences*, Vol. 2, No. 3 (1992), 235–276.
5. See, for example, Kathryn Henderson, 'Flexible sketches and inflexible data bases: Visual communication, conscription devices, and boundary objects in design engineering', *Science, Technology, & Human Values*, Vol. 16, No. 4 (1991), 448–473;

Karin Knorr-Cetina and Klaus Amann, 'Image dissection in natural scientific inquiry', *Science, Technology, & Human Values*, 15, No. 3 (1990), 259–283.

Bibliography

Aldrich, Frances K. and Linda Sheppard, 'Graphicacy: The Fourth 'R'?', *Primary Science Review*, Vol. 64, No. 1 (2000), 8–11.

Andrews, William A., B. Jennifer Andrews, D. A. Balconi and N. J. Purcell, *Discovering Biological Science* (Scarborough, Ontario: Prentice-Hall Canada, 1983).

Barwise, Jon, 'On the circumstantial relation between meaning and content', In Umberto Eco, Marco Santambrogio and Patrizia Violi (eds), *Meaning and Mental Representations* (Bloomington: Indiana University Press, 1988), 23–39.

Bastide, Françoise, 'The iconography of scientific texts: principles of analysis', in Michael Lynch and Steve Woolgar (eds), *Representation in Scientific Practice* (Cambridge, MA: MIT Press, 1990), 187–229.

Bateson, Gregory, *Steps to an Ecology of Mind* (New York: Ballantine, 1972).

Berger, John, *Ways of Seeing* (London: Penguin Books, 1972).

Bertin, Jacques *Semiology of Graphics: Diagrams, Networks, Maps* (Madison, WI: University of Wisconsin Press, 1983).

Bourdieu, Pierre, 'The practice of reflexive sociology (The Paris workshop)', in Pierre Bourdieu and Loïc J. D. Wacquant, *An Invitation to Reflexive Sociology* (Chicago: University of Chicago Press, 1992), 216–260.

Bowen, G. Michael, Wolff-Michael Roth and Michelle K. McGinn, 'Interpretations of graphs by university biology students and practicing scientists: towards a social practice view of scientific representation practices', *Journal of Research in Science Teaching*, Vol. 36, No. 9 (1999), 1020–1043.

Clement, John, 'The concept of variation and misconceptions in Cartesian graphing', *Focus on Learning Problems in Mathematics*, Vol. 11, No. 2 (1989), 77–87.

Damasio, Antonio, *Descartes' Error: Emotion, Reason, and the Human Brain* (New York: HarperCollins, 2000).

Derrida, Jacques, *Limited Inc* (Chicago: University of Chicago Press, 1988).

DiGisi, Lori L. and John B. Willett, 'What high school biology teachers say about their textbook use: A descriptive study', *Journal of Research in Science Teaching*, Vol. 32, No. 1 (1995), 123–142.

Foucault, Michel, *Surveiller et punir: Naissance de la prison* (Paris: Gallimard, 1975).

_____, *This is not a Pipe*, trans. and ed. J. Harkness (Berkeley: University of California Press, 1983).

Gabel, Dorothy L., 'Research on problem solving: Chemistry', in Dorothy L. Gabel (ed), *Handbook of Research on Science Teaching and Learning* (New York: Macmillan Publishing Company), 301–326.

Giroux, Henry A., *Border Crossings: Cultural Workers and the Politics of Education* (New York: Routledge, 1992).

Goodwin, Charles, 'Professional vision', *American Anthropologist*, Vol. 96, No. 4 (1994), 606–633.

Henderson, Kathryn, 'Flexible sketches and inflexible data bases: Visual communication, conscription devices, and boundary objects in design engineering', *Science, Technology, & Human Values*, Vol. 16, No. 4 (1991), 448–473.

Knorr-Cetina, Karin and Klaus Amann, 'Image dissection in natural scientific inquiry', *Science, Technology, & Human Values*, 15, No. 3 (1990), 259–283.

Lakoff, George, *Women, Fire, and Dangerous Things: What Categories Reveal about the Mind* (Chicago: University of Chicago Press, 1987).

Latour, Bruno, *Science in Action: How to Follow Scientists and Engineers through Society* (Milton Keynes: Open University Press, 1987).

_____, *La clef de Berlin et autres leçons d'un amateur de sciences* (Paris: Éditions la Découverte, 1993).

_____, *Pandora's Hope: Essays on the Reality of Science Studies* (Cambridge, MA: Harvard University Press, 1999).

Lave, Jean, *Cognition in Practice: Mind, Mathematics and Culture in Everyday Life* (Cambridge, England: Cambridge University Press, 1988).

Law, John and Michael Lynch, 'Lists, field guides, and the descriptive organization of seeing: Birdwatching as an exemplary observational activity', in Michael Lynch and Steve Woolgar (eds), *Representation in Scientific Practice* (Cambridge, MA: MIT Press, 1990), 267–299.

Leinhardt, Gaea, Orit Zaslavsky and Mary K. Stein, 'Functions, graphs, and graphing: Tasks, learning, and teaching', *Review of Educational Research*, Vol. 60, No. 1 (1990), 1–64.

Lemke, Jay L., 'Multiplying meaning: Visual and verbal semiotics in scientific text', in J. R. Martin & R. Veel (eds), *Reading Science* (London: Routledge, 1998), 87–113.

Livingston, Eric, *An Anthropology of Reading* (Bloomington: Indiana University Press, 1995).

Lynch, Michael, 'Laboratory space and the technological complex: An investigation of topical contextures', in S. Leigh Star (ed), *Ecologies of Knowledge: Work and Poli-*

tics in Science and Technology (Albany, NY: State University of New York Press 1995), 226–256.

_____, 'Method: measurement—ordinary and scientific measurement as ethnomethodological phenomena', in Graham Button (ed), *Ethnomethodology and the Human Sciences* (Cambridge: Cambridge University Press, 1991), 77–108.

Lynch Michael and Steve Woolgar (eds), *Representing as Scientific Practice* (Cambridge, MA: MIT Press, 1990).

Mader, Sylvia S., *Inquiry into Life* 4th ed. (Dubuque, IA: Wm. C. Brown, 1985).

Mayer, Richard E. and Richard B. Anderson, 'The instructive animation: Helping students build connection words and pictures in multimedia learning', *Journal of Educational Psychology*, Vol. 84, No. 4 (1992), 444–452.

Moody, D. E., 'Evolution and the textbook structure of biology', *Science Education*, No. 80, Vol. 4 (1996), 395–418.

Morrison, Ken, 'Some researchable recurrences in discipline-specific inquiry', In D. T. Helm, W. T. Anderson, A. J. Meehan, & Anne W. Rawls (eds), *The Interactional Order: New Directions in the Study of Social Order* (New York: Irvington, 1989), 141–158.

Myers, Greg, 'Every picture tells a story: Illustrations in E. O. Wilson's *Sociobiology*', in Michael Lynch and Steve Woolgar (eds.), *Representation in Scientific Practice* (Cambridge, MA: MIT Press, 1990), 231–265.

Nöth, Winfried, *Handbook of Semiotics* (Bloomington: Indiana University Press, 1990).

Novick, Shimshon and Joseph Nussbaum, 'Junior high school pupils' understanding of the particulate nature of mater: An interview study', *Science Education*, Vol. 62, No. 3 (1978), 273–281.

Ochs, Elinor, *Culture and Language Development: Language Acquisition and Language Socialization in a Samoan Village* (Cambridge: Cambridge University Press, 1988).

Oram, Raymond F., *Biology: Living Systems* 4th ed. (Columbus, OH: Charles E. Merril, 1983).

Piersma, T., J. van Gils and P. de Goeij, 'Holling's functional response model as a tool to link the food-finding mechanism of a probing shorebird with its spatial distribution', *The Journal of Animal Ecology*, Vol. 64 (1995), 493–504.

Pozzer, Lilian L. and Wolff-Michael Roth, 'Prevalence, function, and structure of photographs in high school biology textbooks', *Journal of Research in Science Teaching*, Vol. 40, No. 9 (2002), 1089–1114.

Preece, Jenny and Claude Janvier, 'A study of the interpretation of trends in multiple curve graphs of ecological situations', *School Science and Mathematics*, Vol. 92, No. 6 (1992), 299–306.

Quine, Willard Van Orman, *From Stimulus to Science* (Cambridge, Mass: Harvard University Press, 1995).

Reid, Robin S. and James E. Ellis, 'Impacts of pastoralists on woodlands in South Turkana, Kenya: Livestock mediated tree recruitment', *Ecological Applications*, Vol. 5
(1995), 978–992.

Ricœur, Paul, *From Text to Action: Essays in Hermeneutics, II* trans. K. Blaney and J. B.
Thompson (Evanston, IL: Northwestern University Press, 1991).

Rorty, Richard, *Contingency, Irony, and Solidarity* (Cambridge: Cambridge University
Press, 1989).

Roschelle, Jeremy, 'Learning by collaborating: Convergent conceptual change', *The
Journal of the Learning Sciences*, Vol. 2, No. 3 (1992), 235–276.

Roth, Wolff-Michael, 'Gestures: Their role in teaching and learning', *Review of Educational Research*, Vol. 71, No. 4 (2002), 365–392.

——————————, 'Gesture-speech phenomena, learning and development', *Educational Psychologist*, Vol. 38, No. 4 (2003), 249–263.

——————————, *Toward an Anthropology of Graphing: Semiotic and Anthropological Perspectives* (Dordrecht, The Netherlands: Kluwer Academic Publishers, 2003).

——————————, 'Emergence of graphing practices in scientific research', *Journal of
Cognition and Culture,* (in press).

Roth, Wolff-Michael and G. Michael Bowen, 'Complexities of graphical representations
during lectures: A phenomenological approach', *Learning and Instruction*, Vol. 9,
No. 3 (1999), 235–255.

Roth, Wolff-Michael, G. Michael Bowen and Domenico Masciotra, 'From thing to sign
and "natural object": Toward a genetic phenomenology of graph interpretation', *Science, Technology, & Human Values*, Vol. 27, No. 4 (2002), 327–356.

Roth, Wolff-Michael, G. Michael Bowen and Michelle K. McGinn, 'Differences in
graph-related practices between high school biology textbooks and scientific ecology journals', *Journal of Research in Science Teaching*, Vol. 36, No. 9 (1999), 977–
1019.

Roth, Wolff-Michael and Reinders Duit, 'Emergence, flexibility, and stabilization of language in a physics classroom', *Journal for Research in Science Teaching*, Vol. 40,
No. 9 (2003), 869–897.

Roth, Wolff-Michael and Yew Jin Lee, 'Interpreting unfamiliar graphs: A generative, activity-theoretic model', *Educational Studies in Mathematics* (in press).

Roth, Wolff-Michael, Michelle K. McGinn and G. Michael Bowen, 'How prepared are
preservice teachers to teach scientific inquiry? Levels of performance in scientific
representation practices', *Journal of Science Teacher Education*, Vol. 9, No. 1
(1998), 25–48.

Roth, Wolff-Michael, Kenneth Tobin and Kenneth Shaw 'Cascades of inscriptions and
the re-presentation of nature: How numbers, tables, graphs, and money come to re-

present a rolling ball', *International Journal of Science Education*, Vol. 19, No. 10 (1997), 1075–1091.

Roth, Wolff-Michael, Carolyn Woszczyna and Gilian Smith, 'Affordances and constraints of computers in science education', *Journal of Research in Science Teaching*, Vol. 33, No. 10 (1996), 995–1017.

Schnotz, Wolfgang, E. Picard and A. Hron, 'How do successful and unsuccessful learners use texts and graphics?', *Learning and Instruction*, Vol. 3, No. 3 (1993), 181–199.

Schoultz, Jan, Roger Säljö and Jan Wyndhamn, 'Heavenly talk: Discourse, artifacts, and children's understanding of elementary astronomy', *Human Development*, Vol. 44, No. 1 (2001), 103–118.

St. Martin, Fernande, *Semiotics of Visual Language* (Bloomington: Indiana University Press, 1990).

Suchman, Lucy A. and Brigitte Jordan, 'Interactional troubles in face-to-face survey interviews', *Journal of the American Statistical Association*, Vol. 85 (1990), 232–244.

Tufte, Edward R., *The Visual Display of Quantitative Information* (Cheshire, CT: Graphics Press, 1983).

Ueno, Naoki and N. Arimoto, 'Learning physics by expanding the metacontext of phenomena', *The Quarterly Newsletter of the Laboratory of Comparative Human Cognition*, Vol. 15, No. 1 (1993), 53–63.

Wainer, Howard, 'Understanding graphs and tables', *Educational Researcher*, Vol. 21, No. 1 (1992), 14–23.

Wake, Bruce, 'Best bang for education buck debated at school board', *Peninsula News Review*, 11 December 1998.

Walkerdine, Valerie, *The Mastery of Reason* (London: Routledge, 1988).

_____, 'Redefining the subject in situated cognition theory', in David Kirshner and James A. Whitson (eds), *Situated Cognition* (Mahwah, NJ: Lawrence Erlbaum Associates, 1997), 57–70.

Index

About the authors

 **Wolff-Michael Roth** is Lansdowne Professor of Applied Cognitive Science at the University of Victoria, British Columbia, Canada. For the better part of twelve years, he taught science, mathematics, and computer science at the middle and high school levels. Later, already working at the university, he continued teaching science at the elementary level. His research focuses on various aspects of scientific and mathematical cognition and communication from elementary school to professional practice, including, among others, studies of scientists, technicians, and environmentalists at their work sites. Wolff-Michael Roth teaches doctoral seminars and courses in the analysis of qualitative data. His recent books include *Toward an Anthropology of Science: Semiotic and Activity Theoretic Perspectives, From Articulating Worlds to Talking Science*, and *Rethinking Scientific Literacy* (with A. C. Barton).

 **Lilian Leivas Pozzer-Ardenghi** is a PhD student in Curriculum Studies, at the Faculty of Education, University of Victoria. She received a *licentiate* in Biological Sciences (2000) at the Universidade Federal de Santa Maria, RS, Brazil, and an MA (2003) from the University of Victoria. As part of her Master of Arts program, she investigated the use of photographs in high school science textbooks and in lectures, and investigated students' interpretations of photographs. Her current research interests include gestures in teaching and learning science.

 **JaeYoung Han** is a full-time lecturer of Science Education at Chinju National University of Education in Korea. He studied chemistry education at Seoul National University, where he also received his Ph.D. He visited University of Victoria during the 2003–04 academic year as a postdoctoral researcher. He was a science teacher for four years in ChungDong High School in Seoul. As a member of Teachers for Exciting Science, he is interested in developing science demonstrations for elementary and secondary students. His research interests are diverse, including semiotics in science education, grouping method by students' personalities (thesis), small group learning, environmental education, and learning with analogies, discrepant events, and drawings.

285

Science & Technology Education Library

Series editor: William W. Cobern, *Western Michigan University, Kalamazoo, U.S.A.*

Publications

1. W.-M. Roth: *Authentic School Science.* Knowing and Learning in Open-Inquiry Science Laboratories. 1995   ISBN 0-7923-3088-9; Pb: 0-7923-3307-1
2. L.H. Parker, L.J. Rennie and B.J. Fraser (eds.): *Gender, Science and Mathematics.* Shortening the Shadow. 1996   ISBN 0-7923-3535-X; Pb: 0-7923-3582-1
3. W.-M. Roth: *Designing Communities.* 1997
     ISBN 0-7923-4703-X; Pb: 0-7923-4704-8
4. W.W. Cobern (ed.): *Socio-Cultural Perspectives on Science Education.* An International Dialogue. 1998   ISBN 0-7923-4987-3; Pb: 0-7923-4988-1
5. W.F. McComas (ed.): *The Nature of Science in Science Education.* Rationales and Strategies. 1998   ISBN 0-7923-5080-4
6. J. Gess-Newsome and N.C. Lederman (eds.): *Examining Pedagogical Content Knowledge.* The Construct and its Implications for Science Education. 1999
     ISBN 0-7923-5903-8
7. J. Wallace and W. Louden: *Teacher's Learning.* Stories of Science Education. 2000
     ISBN 0-7923-6259-4; Pb: 0-7923-6260-8
8. D. Shorrocks-Taylor and E.W. Jenkins (eds.): *Learning from Others.* International Comparisons in Education. 2000   ISBN 0-7923-6343-4
9. W.W. Cobern: *Everyday Thoughts about Nature.* A Worldview Investigation of Important Concepts Students Use to Make Sense of Nature with Specific Attention to Science. 2000   ISBN 0-7923-6344-2; Pb: 0-7923-6345-0
10. S.K. Abell (ed.): *Science Teacher Education.* An International Perspective. 2000
     ISBN 0-7923-6455-4
11. K.M. Fisher, J.H. Wandersee and D.E. Moody: *Mapping Biology Knowledge.* 2000
     ISBN 0-7923-6575-5
12. B. Bell and B. Cowie: *Formative Assessment and Science Education.* 2001
     ISBN 0-7923-6768-5; Pb: 0-7923-6769-3
13. D.R. Lavoie and W.-M. Roth (eds.): *Models of Science Teacher Preparation.* Theory into Practice. 2001   ISBN 0-7923-7129-1
14. S.M. Stocklmayer, M.M. Gore and C. Bryant (eds.): *Science Communication in Theory and Practice.* 2001   ISBN 1-4020-0130-4; Pb: 1-4020-0131-2
15. V.J. Mayer (ed.): *Global Science Literacy.* 2002   ISBN 1-4020-0514-8
16. D. Psillos and H. Niedderer (eds.): *Teaching and Learning in the Science Laboratory.* 2002   ISBN 1-4020-1018-4
17. J.K. Gilbert, O. De Jong, R. Justi, D.F. Treagust and J.H. Van Driel (eds.): *Chemical Education: Towards Research-based Practice.* 2003   ISBN 1-4020-1112-1
18. A.E. Lawson: *The Neurological Basis of Learning, Development and Discovery.* Implications for Science and Mathematics Instruction. 2003   ISBN 1-4020-1180-6
19. D.L. Zeidler (ed.): *The Role of Moral Reasoning on Socioscientific Issues and Discourse in Scientific Education.* 2003   ISBN 1-4020-1411-2

Science & Technology Education Library

Series editor: William W. Cobern, *Western Michigan University, Kalamazoo, U.S.A.*

20. P.J. Fensham: *Defining an Identity. The Evolution of Science Education as a Field of Research.* 2003   ISBN 1-4020-1467-8
21. D. Geelan: *Weaving Narrative Nets to Capture Classrooms.* Multimethod Qualitative Approaches for Educational Research. 2003
    ISBN 1-4020-1776-6; Pb: 1-4020-1468-7
22. A. Zohar: *Higher Order Thinking in Science Classrooms: Students' Learning and Teachers' Professional Development.* 2004
    ISBN 1-4020-1852-5; Pb: 1-4020-1853-3
23. C.S. Wallace, B. Hand, V. Prain: *Writing and Learning in the Science Classroom.* 2004   ISBN 1-4020-2017-1
24. I.A. Halloun: *Modeling Theory in Science Education.* 2004   ISBN 1-4020-2139-9
25. L.B. Flick and N.G. Lederman (eds.): *Scientific Inquiry and the Nature of Science.* Implications for Teaching, Learning, and Teacher Education. 2004
    ISBN 1-4020-2671-4
26. W.-M. Roth, L. Pozzer-Ardenghi and J.Y. Han: *Critical Graphicacy.* Understanding Visual Representation Practices in School Science. 2005   ISBN 1-4020-3375-3
27. M.J. de Vries: *Teaching about Technology.* An Introduction to the Philosophy of Technology for Non-philosophers. 2005   ISBN 1-4020-3409-1
28. R. Nola and G. Irzik: *Philosophy, Science, Education and Culture.* 2005
    ISBN 1-4020-3769-4
29. S. Alsop (ed.): *Beyond Cartesian Dualism.* Encountering Affect in the Teaching and Learning of Science. 2005   ISBN 1-4020-3807-0